LOCOMOTIVES

of the

LMS NCC

and its predecessors

Dedicated to Frank Dunlop of Coleraine, whose knowledge of NCC engines has enriched this book and helped keep steam alive on Irish railways by the performances of WT 2-6-4T No 4.

Frank on his last day in railway service, 1990.
Joe Cassells

First Edition
First impression

© WT Scott and Colourpoint Books 2008

Designed by Colourpoint Books, Newtownards
Printed by W&G Baird Ltd

ISBN 978 1 904242 84 0

Colourpoint Books
Colourpoint House
Jubilee Business Park
21 Jubilee Road
NEWTOWNARDS
County Down
Northern Ireland
BT23 4YH
Tel: 028 9182 6339
Fax: 028 9182 1900
E-mail: info@colourpoint.co.uk
Web site: www.colourpoint.co.uk

WT Scott has been a lifelong railway enthusiast and in particular a student of NCC locomotives. He came of a railway family, the grandson of a BNCR stationmaster. He has written several articles for various railway magazines on Irish locomotive design and performance. He has also built several models of BNCR and GS&WR locomotives, which ran on the layout of the late Drew Donaldson. Apart from the study of steam locomotives of many countries, his main interests are golf, rugby and cricket.

Cover pictures

Front: Resplendent in LMS(NCC) crimson lake, W class 2-6-0 No 94 *The Maine* prepares to leave Portrush with a train for Belfast around 1936.
A painting by Jack Hill (by kind permission of the Lord O'Neill)

Rear upper: U2 class 4-4-0 No 74 *Dunluce Castle* in post-war LMS(NCC) black livery at York Road in 1950.
WP De Beer, ColourRail

Rear lower: U2 class 4-4-0 No 74 at York Road in 1962 after restoration to NCC red livery for the Belfast Transport Museum. It is now at Cultra.
Des FitzGerald

End Papers

Front: W class 2-6-0 No 102 appoaching Ballymoney on 3 July 1953 with the down Royal Train from Lisburn to Lisahally.
Kenneth Bennington

Rear: The unnamed 'Scotch engine', No 77, between the tunnels at Downhill about 1948 with a Coleraine to Londonderry train.
Kenneth Bennington

LOCOMOTIVES
of the
LMS NCC
and its predecessors

WILLIAM SCOTT

COLOURPOINT

The first superheated locomotives on the NCC were U class 4-4-0s Nos 69 and 70. This is the official photograph of No 70 after completion at Derby in July 1914. She is in the Midland livery of that era with a small crest on the splasher. She is an interesting amalgamation of Derby styling and York Road engineering.

Official NCC photograph

Contents

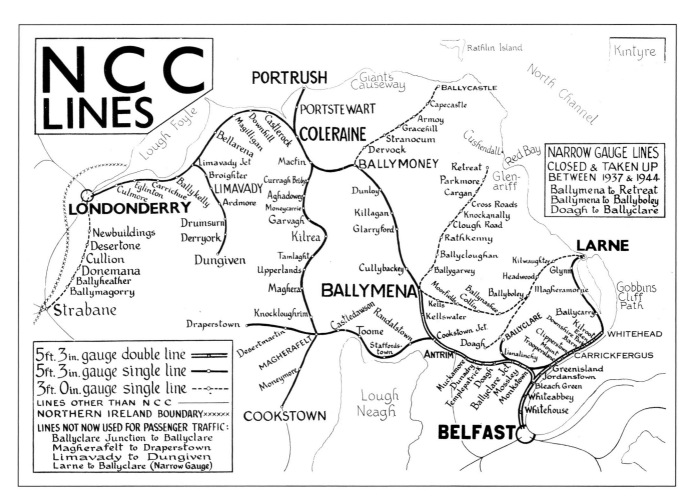

The NCC footplate, in this case the controls of W class 2-6-0 No 95 *The Braid*, when almost new in 1936, but old enough for the crew to had added unofficial padding to the steam valve on the right hand injector.

OS Nock, courtesy CP Friel

Preface and Acknowledgements

The origins of this book go back over forty years and came from a railway connection. My grandfather was successively porter and stationmaster at Coleraine, Downhill and Aghadowey. My Uncle Willie, who was a clerk on the Northern Counties before the First World War, introduced me to the NCC in the early 1950s.

I am indebted to a group of men, some engineers and some in the Operating Department, whose interest in railways went far beyond merely the work at which they earned their salaries. Harold Houston joined the NCC in 1920 and served for over 50 years as an engineer. My first acquaintance was through reading his papers on NCC Locomotive Development in successive issues of the IRRS Journal. Later, I got to know him personally – soon learning that he was only happy to travel in the same compartment as a train timer if the window was kept shut! I asked him one day for some information on BNCR engines, and a few days later an envelope arrived in the post with precise details of the BNCR and NCC compounds. WAG (Billy) McAfee, also an NCC-trained engineer, has contributed to the literature of Irish locomotive development, and was helpful to me over many years. In addition, the detailed work of the famous RN Clements was a valuable guide. What these men wrote, and what they shared with me were, over many years, the real foundation of this book.

I am indebted to Philip Atkins, formerly of the National Railway Museum, York, for some invaluable insights into the development of the 2-6-4T concept at Derby prior to 1927 and the NCC design from 1944.

On the operational side, I have to thank the late Locomotive Inspector Billy Hanley for instructing me, unofficially, in the art of firing and driving. Although he described himself as just 'a labouring man', Hanley was by instinct a first-rate and entirely self-taught engineer whose personal memories and copious diary notes enabled me to include so many stories and facts from the first four decades of the twentieth century.

Hanley's successor as Locomotive Inspector was Frank Dunlop whom I count as a lifelong friend. Frank's supreme talent was getting things done: problems on the railway were there to be solved, with as little fuss as possible. He was in his element on big occasions like the annual Apprentice Boys' Day at Derry, or a busy summer Saturday at Portrush, where there seemed to be too many specials and not enough engines. To see him on these days, leading by example and giving a steady stream of help, instruction, advice and encouragement, was to watch a consummate professional at work. What no-one saw was the hours of preparation Frank put into these occasions,

rostering the right men, choosing the best engines and arranging for everything from a wagon of coal down to a spare vacuum bag to be previously supplied. Latterly Frank was an indispensable source of help and advice to the Railway Preservation Society of Ireland. One of the last great triumphs of his career was the organisation of an RPSI private charter to the Permanent Way Institute when they visited Ireland in 1985. That day, involving seven drivers, two engines, 336 miles of steam running and a seventeen hour day for Frank, could well rank as the climax of his career. It is with much pleasure that I dedicate this book to him.

The men who drove and fired the engines are too many to mention individually, but few NCC men could not find time for a word with an enthusiast, or an invitation to the footplate when authority was not looking. Sadly, there are now few survivors of the age of steam, and this book is a small tribute to their memory.

I have received much help from fellow enthusiasts. Joe Cassells has turned my execrable handwriting into a typed manuscript, and made many helpful suggestions and additions, particularly to the chapters on the moguls and mogul Tanks. Irwin Pryce, a lifelong Great Northern enthusiast who has studied the design and performance of the tank engines for over forty years, also supplied much information, and an invaluable analysis of preserved NCC No 4 from an operational and maintenance point of view. Peter Scott MBE who, as RPSI locomotive officer, shares with Irwin an intimate acquaintance with No 4, has provided some of his own insights as to how the design of the mogul Tanks might have been improved even further. Charles Friel has been exceptionally helpful in providing photographs from his collection and proof reading the manuscript. Nelson Poots provided many railway reminiscences from his home town of Larne. Ian Wilson generously assisted with proof reading, whilst Des FitzGerald, Kenneth Bennington and Derek Young loaned photographs. Michael McMahon also assisted me with valuable information.

I must also thank Norman Johnston and the staff of Colourpoint. Since its inception, Norman has taken a close personal interest in this book. He has willingly shared the results of his own researches into the development of Irish steam locomotives, and this book has benefited greatly from his careful and rigorous analysis. No author could wish for a more knowledgeable or sympathetic publisher, and he has saved me from many errors. Any which remain – and I trust they will be few - are my responsibility alone.

Introduction

Before its acquisition by the Midland Railway of England in 1903, the Belfast and Northern Counties Railway had a route mileage of 201 on the 5'3" gauge and 48½ on the 3'0" gauge, making it the fourth largest railway in Ireland. In 1924 the Ballycastle Railway joined the group, which the previous year had become the London Midland and Scottish Railway (Northern Counties Committee), bringing the 3'0" gauge mileage to 64½. Like most railways, the BNCR was built up piecemeal. The principal components were the Belfast and Ballymena Railway, opened in 1848, the Londonderry and Coleraine Railway, opened in 1852–53, and the Ballymena, Ballymoney Coleraine and Portrush Junction Railway, opened in 1855. The BNCR was a prosperous and well managed railway running short trains at moderate speeds. Things continued thus throughout the Midland period, and into the early 1930s, when, in a few years, the LMS transformed the NCC into a railway running some of the fastest trains in Ireland. Its locomotive stock never acquired the uniformity of appearance of the GNRI, but the fleet's very diversity attracted locomotive enthusiasts from all over the British Isles. In Coleraine, for example, an outside framed 2-4-0, a two cylinder compound and a modern 2-6-0 might be seen, together with a 4-4-0 of unmistakeably Derby outline. This variety, which did not survive the Second World War, made the NCC a much visited railway in the 1930s.

One facet of NCC operation which was without equal among Irish companies was its single track main line sections. Loops were realigned and signalling set to give a fast line through each station for trains in either direction. This, combined with the excellent Manson tablet catcher, meant that tokens could be exchanged at speeds in excess of 60mph. There were few greater thrills than standing on the footplate of a mogul as she swept through Dunloy to the accompaniment of the resounding crash as the tablets were exchanged. Lineside catcher arms were kept well aligned and the exchange was rarely missed. When it was, there could be a long delay as the engine crew searched the long grass at the lineside, or sometimes the guttering of the station roof, for the errant token!

After a brief period of ownership by British Railways from 1 January 1948, under the name of the Railway Executive (NCC), the dead hand of the Ulster Transport Authority took over on 1 April 1949, but although the NCC's magic name was gone, the tradition of clean locomotives and smart running persisted almost until the end of steam. Locomen rarely spoke of the UTA: even after twenty years men often referred to their railway as 'The Midland' and always described their engines as 'NCC'.

WT class 2-6-4T No 54, with an up train, passing a remotely controlled somersault signal at Ballykelly halt in 1954. The wooden housing contained the electric motor. Ballykelly cabin controlled the runway at the nearby RAF base. Conventional signal wires could not be used, as the line crossed the runway!

Author

Chapter 1
Pre-BNCR Days

Since the BNCR was the Belfast and Ballymena Railway under a new name, we can trace its locomotive history back continuously to 1847. Before following the main theme therefore, it makes sense to deal firstly with the locomotives of the two smaller companies the BNCR absorbed or leased in 1861.

The Londonderry and Coleraine Railway.

The Londonderry and Coleraine Railway was incorporated in 1845 and opened in stages in 1852–53. In its nine years of independent operation, this railway acquired seven new and several second-hand engines of varying degrees of uselessness. The new engines consisted of five 2-2-0 well tanks, one 0-4-2 and one 2-4-0. No numerical list of L&CR locomotives before 1861 has survived and piecing together the numbers carried by the L&CR fleet has been complicated by exchanges of engines between the L&CR and the neighbouring Londonderry and Enniskillen.

Records are sketchy and incomplete, but it seems that in 1845 an order was placed with Longridge for three long-boiler engines, to Robert Stephenson's specification, two later being cancelled due to the delays in constructing the line. The one which actually arrived was an inside cylinder 2-4-0 goods engine, which lay in Londonderry for about two years from 1846, before being loaned to the Londonderry and Enniskillen Railway, a sister company on the other bank of the Foyle, which shared some directors and had placed an identical order with Longridge. This line had opened its first section in 1847. This first engine

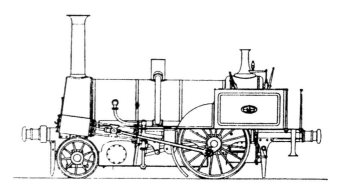

One of the Sharp 2-2-0WTs built for the Londonderry and Coleraine Railway in 1853. *The Locomotive*

probably never received an L&CR number, but may have carried L&ER No 3.

A different Longridge 2-4-0 came back to the L&CR in February 1851 (or possibly February 1852. This 2-4-0 was definitely not the same one, as it had outside cylinders and larger driving wheels. Most likely this engine was L&CR No 2 and in 1861 it became BNCR No 27. The reason it is thought to have been No 2 is that the L&CR had another mystery engine, purchased in 1848, which was withdrawn in 1858 and was probably No 1.

The next order was for six small 2-2-0 well tank engines of the NB Adams patent, built by Sharp Stewart as their Works Nos 716–19, 722, and 723. However,

The BNCR started numbering their engines with the B&BR fleet which went up to 19, then the BBC&PJR up to 25 and the L&CR to 33. The last L&CR engine was a Fairbairn 2-4-0 which arrived in 1861, too late to count as L&CR stock and so was taken straight into the BNCR list as No 34. She was photographed here at Belfast about 1898 and was scrapped in 1901. Fairbairns were prolific builders in the middle of the 19th century and supplied 63 engines to Ireland leaving them fifth behind Beyer Peacock, Sharp Stewart, Neilson Reid and Vulcan Foundry.
Author's collection

one, No 719, was sent to Dublin just after delivery in April 1853 and exhibited by the makers at the Dublin International Exhibition at Leinster House, where it attracted the interest of the directors of the Newry and Enniskillen Railway. They made an arrangement to buy it, so when it was returned in December 1853, it was transported to Newry and never ran on the L&CR. The other five became either L&CR Nos 2–6 or 3–7, assuming that the exhibition engine never carried a L&CR number, but if the six were numbered, they could have been 2–7 or 3–8.

The well-tanks became BNCR Nos 28–32. The late Harold Houston said that these engines appeared in BNCR records as 2-2-2 well tanks and that the trailing wheels were fitted to cure their unsteadiness at speed. One, No 4, was derailed on 14 March 1855, at Rosses Bay curve, a couple of miles from Londonderry, whilst working the mail train, and rolled over, killing the driver. Their outside cylinders would not have helped their stability. Harold Houston thought them quite unsuitable for a 30 mile line, but they were chosen for their light axle loading, and for their cheapness – they cost only £1130 each.

However, the only one known for certain to have been a 2-2-2WT is No 28, sold to the contractor Greg in 1871. No 29 was rebuilt as a light 2-4-0 in 1869, and No 32 was withdrawn that year. An 1871 Engineer's Report mentions two four-wheel engines in stock, likely to have been Nos 30 and 31, but it is possible (though unlikely) that they became 2-2-2WTs between that date and withdrawal in 1880 and 1878 respectively.

Two further engines were added in 1858-59. The first was a second-hand 2-2-2 well tank built by Fairbairn in 1855 for the Ballymena, Ballymoney, Coleraine and Portrush Junction Railway (BBC&PJR) and sold to the L&CR some time between April 1858 and January 1859. Its L&CR number is uncertain, but it possibly replaced the original, mystery, No 1 and took the same number or, alternatively was No 7. It became BNCR No 26, which suggests it was No 1 on the L&CR, though it may have received number 26 to put it beside Nos 24 and 25 of the same type.

The second was a Grendon 0-4-2 built in January 1859. It was definitely L&CR No 8 and became BNCR No 33. Finally, the L&CR had one 2-4-0 engine on order from Fairbairn in 1860, which arrived the following year, after the BNCR takeover and became BNCR No 34. Further details are in Tables 1 and 7 (pages 20 & 22).

In all the L&CR contribution to the BNCR was, therefore, nine locomotives. The Company's main claim to locomotive fame was that for one year, 1853-1854, Robert Fairlie was its Locomotive Superintendent. One wonders whether the idea of the 'Double Fairlie' might have been inspired by the fairly common sight of a pair of 2-2-0 well tanks with open cabs running back to back. With the exception of No 34, which after rebuilding lasted until 1901, all the L&CR engines had gone by 1890 and had no influence on subsequent BNCR locomotive policy.

The Ballymena, Ballymoney, Coleraine and Portrush Junction Railway

The BBC&PJR's locomotives were purchased on the advice of William Dargan, the contractor who built the line. Despite its long name, this concern, which opened in December 1855, had only 40 miles of track and only seven locomotives of two classes. Its records,

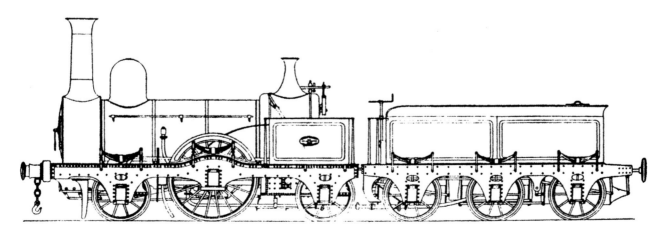

A standard Sharp Stewart 2-2-2 of the type built for the BBC&PJR. Nos 1–4 were built in 1855 and became BNCR Nos 20–23. They had 5'6" driving wheels and 15" x 20" cylinders but their boilers were pressed to 120psi, making them more powerful than the Sharps supplied to the Belfast and Ballymena Railway in 1847.

JH Houston, IRRS collection

The only known photographs of BBC&PJR engines are this one and another of 2-2-2WT No 24, taken about 1895. She has lasted long enough to get the vacuum brake. Indeed her leading and trailing wheels are braked as well. She was built by Fairbairn in 1855.

Author's collection

however, were in better order than the L&CR's, and the historian's task is correspondingly easier. The first four engines (2-2-2 Nos 1–4) were typical Sharp Stewart products, while the last three (Nos 5–7) were Fairbairn 2-2-2 well tanks, though only two went directly to the BNCR. They became BNCR Nos 20–23 and 24–25 respectively. Further details are in Tables 2 and 6.

The third well tank was sold to the L&CR in 1858, as explained above. The sale of this engine might indicate that the directors reckoned they had made a mistake in buying tank engines rather than goods engines. Since trains ran through to Portrush before 1860 the question arises as to why they needed tank engines at all.

Of the tender engines, No 21 was rebuilt in 1870, with a new boiler, but remained a single. Nos 20, 22 and 23 were rebuilt with new boilers as 2-4-0s in 1870–71, later becoming Class H. No 22 was scrapped in 1877, following damage in the Moylena collision in December 1876 and No 23 in 1886. No 21 survived until 1893 and was the last 2-2-2 to run on the NCC. No 20 was the last BBC&PJR engine to run, and the only one to survive into Midland ownership. It was scrapped in 1906.

Of the tank engines, No 25 was rebuilt as a small 2-4-0 tender engine in 1867 and withdrawn in 1883. No 24 was the last Fairbairn to run on the NCC, as well as the last single driver. As was the case with the L&CR, the company's locomotive policy had no influence on that of the BNCR after 1860.

The Belfast and Ballymena Railway

As senior member and most influential component of the BNCR, the Belfast and Ballymena Railway and its locomotives dictated the early policy of the fledgling BNCR.

The first locomotive to work the line was *Spitfire*, bought second hand from the Ulster Railway in 1847. It was bought by the contractor, William Dargan, in September 1847, because an engine was needed for ballasting the line and none of the B&BR's own locomotives had arrived. The deal was that the engine would be repurchased by the Ulster Railway after the ballasting, at the same price, less a deduction for wear and tear.

Spitfire had been built as a double-framed 2-2-2 tender locomotive by Sharp Roberts (maker's No 57) in 1839, but was rebuilt as a 2-2-2WT, possibly as early as 1842. It was converted from the Ulster's 6'2" gauge to the standard Irish 5'3" (probably at the Belfast foundry of Coates and Young) before being handed over. The cylinders (originally 14" in diameter) were reduced to 11", then or earlier.

There is a further twist to the story. In December 1847, when the time came to return *Spitfire*, it had been so badly damaged that it could not be returned. In its place, the Ulster Railway agreed to accept *Hawk*, a new Sharp Brothers 2-2-2, and to pay the Belfast and Ballymena £300, the difference in value between the two engines. *Hawk* became UR No 13. According to Harold Houston, this exchange took place in October 1849, two years later. Clements regards this as unlikely, as the £300 payment was made on 23 December 1847. As part of the transaction, the two engines swopped names. However, the B&BR seem not to have attached the *Spitfire* nameplates to the UR engine, as the UR records don't refer to No 13 as *Spitfire* before 1855.

The little tank engine subsequently led an eventful life. She became B&BR No 1 *Hawk* and was rebuilt again in October 1854 as a 2-4-0WT, with smaller 5'6" driving wheels. *Hawk* was withdrawn in 1863 and sold privately to Thomas Firth, Locomotive Engineer of the Belfast, Holywood and Bangor Railway, for £366.

Years ago it was thought that she was the mystery 'No 3' on the BH&BR, but the late RN Clements debunked this theory by showing that Firth actually sold *Hawk* almost immediately to John Killeen, contractor on the Kilmesan to Athboy section of the Dublin and Meath Railway. He in turn advertised it for sale at Trim, Co Meath, later that year and it was bought by contractors French and Cheyne, for use on the Midland Counties and Shannon Junction Railway.

When that contract was suspended, it was sent for repair by the Midland Great Western Railway at their Broadstone works, but the owners did not pay the bill, and it lay at Broadstone for two years until it was finally sold for scrap by the MGWR in 1872. Regardless of the precise details of her career, *Hawk* in her lifetime

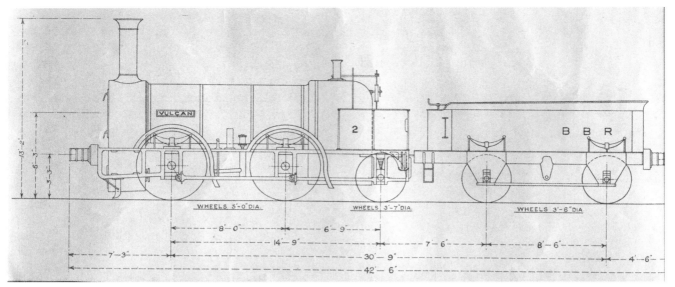

Bury 0-4-2 No 2 *Vulcan*, built in 1847 and withdrawn in 1869.

JH Houston

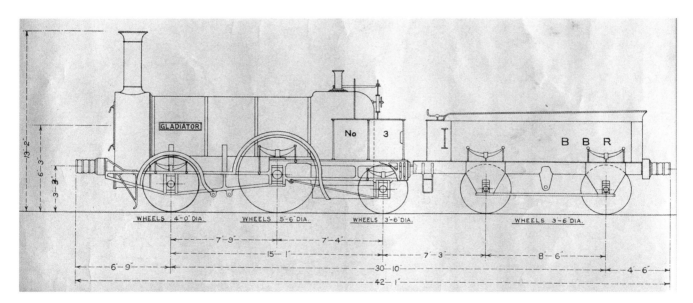

Bury 2-2-2 No 3 *Gladiator*, built in 1847, rebuilt as a 2-4-0 in 1855 and withdrawn in 1868. Nos 4–6 were identical.

JH Houston

certainly covered a fascinating selection of routes!

The Bury, Curtis and Kennedy engines

Concurrently with this saga, the Belfast and Ballymena ordered five new engines from Bury, Curtis and Kennedy, which were delivered in October 1847. These were typical Bury engines with bar frames and a 'hay stack' firebox. An example of this type can be seen plinthed at Cork station. Four of the engines were 2-2-2 tender engines and the other was a 0-4-2. Bury, Curtis and Kennedy were important builders for the early Irish railways, supplying 42 engines, mostly 2-2-2s. The Belfast and Ballymena engines carried only

names at first, numbers not appearing until 1852.

The 0-4-2 was No 2 *Vulcan*. She was intended for goods and ballast work and cost £1950. The 2-2-2s were Nos 3 *Gladiator*, 4 *Hercules*, 5 *Queen* and 6 *Prince*. The four 2-2-2s were for passenger work, and cost £1912 each. Dimensionally, they were engines whose size was typical of their period. Until 1845 Bury had resisted the move to six-wheeled engines, preferring 2-2-0s and 0-4-0s. In 1848 the B&BR asked for link motion to be fitted to the engines, as the original gear was unsatisfactory. This was probably some form of gab gear. Increasing train weights and speeds overtook the tiny Bury engines, and three were rebuilt as 2-4-0s

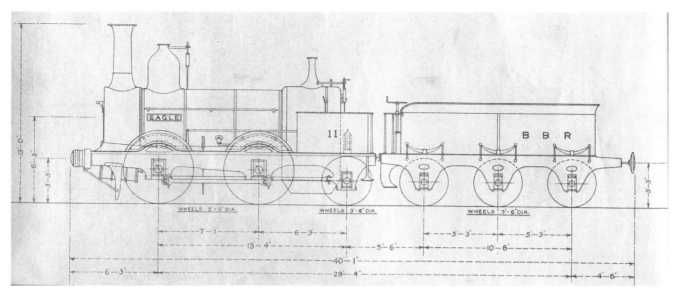

Sharp 0-4-2 No 11 *Eagle*, built in 1847 and withdrawn in 1873.

JH Houston

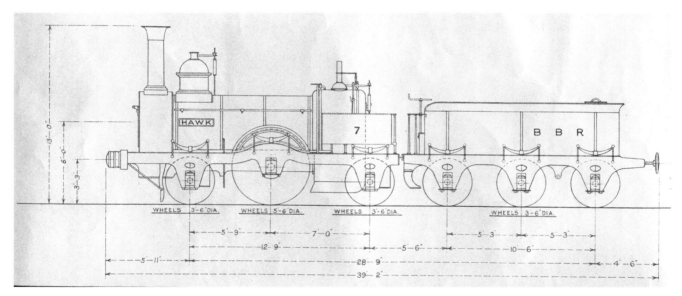

Sharp 2-2-2 No 7 *Hawk*, built in 1847, and transferred to the Ulster Railway in 1847 or 1849. Nos 8–10 were identical.

JH Houston

in 1853–55, the exception being No 6 which finished her days as a 2-2-2. The later history of these engines is summarised in Table 3, page 20.

The Locomotive Superintendent in 1860 was Alexander Yorston, who served both the B&BR and the BNCR from 1849 to 1868. His report on the locomotive stock in 1860 makes interesting reading. He described all five of the Bury, Curtis and Kennedy engines as good, considering their age, and all still working at their original pressure of 80psi, but he recommended the disposal of *Hawk*. His report concluded that "the other engines are in first rate order". This sentence refers to the 13 Sharp Brothers and Sharp, Stewart

engines which followed the Bury brood.

Between 1860 and 1868 something went wrong at York Road and in April 1868 Yorston had to report that the engines were in a poor state. Bury engines Nos 2, 4, 5 and 6 were especially bad, and Nos 27, 28, 29, 30, 31 and 32 were in an equally parlous state – not surprising as these engines formed almost the entire stock of the Londonderry & Coleraine Railway. The directors took drastic action, sending No 3, the best of the Bury engines, to Grendons in 1868, and scrapping the rest as soon as possible. Yorston, who was now an old man, was eased out of office and replaced by Edward Leigh of the Newry and Armagh Railway, from 1868–75.

The specification of No 3 as a 'rebuild' is interesting. The scheme included brass tubes and dome, plates to be of Lowmoor iron, new cylinders, pistons and valve rods and new frames. (Brass seems a peculiar material for tubes which had to be expanded, but the type of brass used contained as much as 75% copper. This made it malleable, without being as soft as copper.) Apart from the wheels and axles, there seemed to be very little left of the original engine! Although classed by the BNCR as a rebuild, she must surely have counted as a new engine built by Grendons. Estimating the extent of rebuilding and assembling carried out at York Road is a recurring problem in BNCR and NCC locomotive history.

The Sharp engines

Like most pioneering railways, the Belfast and Ballymena shopped around for its engines and, as well as the Bury, Curtis engines, they ordered five new locomotives from Sharp Brothers in 1847 – four singles and a 0-4-2. These were built to orders 165 and 155, and a larger 0-4-2 came in 1851 to order 674. Details of these engines are in Table 3, page 20.

The Sharp singles were Nos 7–10, As recorded earlier, No 7 became UR No 13 *Spitfire*. Nos 8–10 were never rebuilt, finishing their days as 2-2-2s. At the time of their withdrawal in the 1870s it would have been only slightly more expensive to replace them with new engines than to rebuild them.

The first of the 0-4-2s, No 11 *Eagle*, worked main line goods trains, but suffered a boiler explosion in 1857 which may have been caused by the crew tampering with the safety valves to extract a little more power from the engine. *Eagle* was reported at the time to have been nearly destroyed, so her repair must have been virtually a rebuild.

The second Sharp 0-4-2, No 7 *Ostrich*, which took the number of the earlier single, was the subject of a dispute between Sharps and the B&BR due to late delivery of the engine. It had been promised for June 1850, but had still not arrived by November. The B&BR then refused to accept the engine, and after some haggling Sharps offered a reduction of £260 – about 12% of the original cost. On this basis a deal was struck, the B&BR reckoning that it represented the value of the traffic they had lost due to the engine not being available. No 7 finally arrived in December 1850 and joined No 11 on the main line goods turns until "getting into bad order". This phrase is a recurrent one in reference to the early BNCR locomotives, and indicates that maintenance at York Road became sketchy in the last days of the aged Yorston. It did not help, either, that relatively small engines were being flogged to death as traffic increased.

Table 5, page 21, gives the dimensions of the early Sharp engines. The use of brass for tubes may come as something of a surprise to preservationists in the 21st century, but in the 1850s both Sharps and Grendons used it frequently, for all that it

The NCC in the mid 20th century was often accused of pulling goods trains with four-coupled engines and passengers with six-coupled, but the practice has a long history and the picture shows H class 2-4-0 No 16 at Whitehead with a goods train for Belfast. The crew have only a weatherboard for protection. The engine's chimney must surely test the loading gauge and the inside cylinders are clearly shown. There is no front coupling either, a practice carried over onto the NCC.
The driver, well-weathered and bearded, with white shirt, tie and gold watch chain, wears the company cap. The fireman has a grease-top cap. Both are watching the camera but the engine knows the road anyway. The wagon is typical BNCR with Wilson type axle boxes.

CP Friel collection

seems an unsuitable metal for expansion.

1856 was a watershed year for the B&BR. Increasing train weights showed the need for engines with coupled driving wheels, and no single-drivers were ordered after 1847. The next order placed with Sharps (by now known as Sharp, Stewart) was for six 2-4-0 engines and tenders. They become known as the H class, and a chronology of the class is contained in Table 4. These double-framed 2-4-0s were superb engines and served well into MR(NCC) days, though after their many rebuildings little of the originals can have remained except the wheel centres. When they arrived the engines had weather boards instead of cabs, and smokebox wing plates in the fashion of the time. Brass beading was fitted around the splashers, and the boilers carried brass domes on the front ring. The original boiler was

smaller in diameter than the firebox, the front of which was covered with a burnished brass ring. Engines 15 and 16 became regulars on the Dungiven branch, and had weather boards on the tender for this reason.

The basic dimensions of the class are contained in Table 6, page 22.

The last two engines ordered by the B&BR were Sharp Stewart 0-6-0s intended for goods trains and known later as the L class. Two were delivered in 1857 at a cost of £2,580 each. A third, very similar, engine was ordered from Sharp Stewart in December 1860 and, although it never saw service with the B&BR, it may conveniently be considered here. Nos 18 and 19 (Sharp Stewart 993 and 994) were withdrawn respectively in 1925 and 1933. No 18 was rebuilt in 1889 with 17"x24" cylinders and a new boiler, and reboilered again in 1908. No 19 was

Continued on page 17

This picture shows No 14 at Londonderry in 1902. The gadget on the weatherboard looks like it might be a brake handle. Both this picture and the previous one show the engines after being reboilered in the mid-1870s with the dome on the middle ring. Boilers in this period usually had a firebox casing of larger diameter than the boiler.

Locomotive Publishing Co

In the 1890s, the H class were again rebuilt and at last received cabs. No 12 is seen at Londonderry in September 1898. Trailing views can be useful, as in this case, where the picture shows how cramped the cab was, with the fire box protruding well into the cab space. The crew are working the footplate while a cleaner is putting the finishing touches to an already immaculate engine. The chimney has a copper cap.

Ken Nunn Collection, LCGB

Left: Nos 12–15 received a final rebuild in 1908–13, giving them 4'0" boilers with flush firebox casings. No 13 is seen at Ballymoney in 1920. At this stage she was based in Coleraine and would have been sent to Ballymoney for shunting duties. The engine behind is K class 0-6-0 No 31 which is waiting to leave on a goods for Belfast.
No 13 was built in 1856, rebuilt in 1878, 1896 and 1908 and withdrawn in 1924.
Ken Nunn Collection, LCGB

Left: H class No 15, photographed at Portrush with the large wooden goods shed as a background – this was a traditional spot for photographing NCC engines. She has still got her brass dome and, like No 16, opposite, has a weather board on the tender for branch line working. Note the very tiny balance weights on the driving wheels.

Author's collection

(Continued from page 15)

rebuilt in 1888 with a new boiler and 17"x24" cylinders, and fitted with a 'five foot' boiler in 1920. At the time of her withdrawal in 1933 she was the oldest engine on the NCC, with a working life of 76 years. The third engine, whose life began as Belfast and Northern Counties No 35 became that company's first new engine. She was rebuilt in 1883 with a new boiler and 17"x24" cylinders and again in 1902 with a 'five foot' boiler. Dimensions of the L class 0-6-0s appear in Table 6.

The so called 'five foot boiler' is mentioned here for the first time and No 35 was the first to receive one. This boiler was actually 4'8¼" in diameter, but with cladding was close enough to five feet. There were 193 small tubes but no superheater. It was used mainly for goods locomotives but a 9'10" long version was given to some engines of classes B, C, F and G, as well as to the solitary M class 0-4-2.

Left: L class 0-6-0 No 35 arrived in 1861, just too late to be included in the B&BR List. She displays a common feature of 0-6-0s of this period – a sloping smokebox front, the angle being determined by the slope of the cylinders. The dome is placed on the forward ring of the boiler, a common practice in the 1860s. A vacuum brake has been fitted at some later time. The L Class had small 4'7" driving wheels, hence the coupling rod splashers.

Real Photographs Ltd

Opposite centre: H class No 12 in immaculate condition at Belfast in 1913. She has just been rebuilt and has been given a continuous handrail and a slightly higher pitched boiler than No 13 on the previous page. Even the frames are lined out and she has retained her brass dome. Note how the rod to work the sanding is elegantly curved round the brass beaded splasher – a real Sharp gem some 60 years after she was built. Mechanically she is not quite so sound, the brake blocks being well worn on both leading and trailing wheels.

Official NCC photograph

Opposite bottom: Nos 16 and 17 did not receive parallel boilers and the latter was replaced in 1908. Here, No 16 is seen in her final condition with 1898 boiler and tender weatherboard for working the Dungiven branch. The strange apparatus on the cab side is still there, but what catches the eye now is the tablet exchanger. This lethal looking apparatus is mounted on the tender beside the weatherboard. To remove the tablet, the fireman will have to stand in the gap between engine and tender (no cab doors) and reach out into space. Health and Safety and the Nanny State would produce a whole 'litter of kittens' at such an idea.

Author's collection

This is the same locomotive after its second rebuild in 1902, when it was the first BNCR engine to receive a 'Five Foot' boiler. The engine has no back sanding but has acquired steam sanding to the front drivers. The livery appears to be BNCR invisible green but the highly polished brass dome has almost vanished.

Harold Fayle

The first two members of the L class were Nos 18 and 19, built for the B&BR by Sharp Stewart in 1857. Originally, they resembled No 35 in the first view but were rebuilt in 1888–89 and again in 1908. No 19 was a surprising choice for a more drastic rebuild in 1920 when it received a new 5'0" boiler with large firebox, as used on the K1 rebuilds which followed. At this time it was already 63 years old!

Locomotive Publishing Co

This is a three-quarter front view of No 19 in Belfast yard in July 1932 towards the end of its working life. It was withdrawn in December 1933 at the venerable age of 77.

JAGH Coltas

The BNCR's locomotive position at the amalgamation of 1861.

The Belfast and Northern Counties Railway started life with 33 locomotives, 19 of which came from the B&BR, six from the BBC&PJR and eight from the L&CR. Only one engine from each of the last two companies survived into the twentieth century, but eight of the B&BR locomotives received MR(NCC) numbers.

The Belfast and Ballymena engines were numbered 1 to 19, the Ballymena & Ballymoney engines from 20 to 25, and the Londonderry and Coleraine from

26 to 33. The B&BR naming policy finished with *Ostrich* in 1851, after which the engines were given numbers. In relation to builders, in 1861 the BNCR owned 23 Sharp Stewart engines, five from Bury, three from Fairbairn, and one each from Longridge and Grendon. Two other engines were under construction at the time of the amalgamation. The Fairbairn 2-4-0 ordered by the L&CR and the B&BR's Sharp Stewart 0-6-0 arrived in 1861 and were allocated numbers 34 and 35.

The dominance of Sharp as a supplier of motive power is clear. Nearly 70% of the BNCR engine stock in 1861 came from this maker. However, on the Dublin and Belfast Junction Railway, Sharps supplied *all* of the early engines, whilst the corresponding figures for 1861 on the Ulster Railway were 82% and for the Dublin and Drogheda Railway, 55%. This trend was rapidly reversed in the 1870s, and the BNCR bought its last Sharp Stewart engine in 1876.

One reason for the decline was the decision by locomotive designer Charles Beyer to leave Sharp Stewart in 1854 and go into partnership with Richard Peacock. The firm that they founded carried, of course, one of the most celebrated names in the locomotive building industry.

Sharps, in their earlier period, had manufactured textile machinery, and young Charles Beyer of Dresden came to England to study English textile machinery and report back to Dresden. He returned in 1834 and was employed as a junior draughtsman with Sharps, who were now putting their main efforts into the profitable field of locomotives. Beyer's genius was soon recognised, and soon all the Sharp locomotives originated on his drawing board. Sharp-Beyer engines were characterised by sandwich outside and plate inside frames, and overall sturdiness of construction, and they were long-lasting, though rather heavier than average. When Beyer left Sharps he took many of their customers with him, and certainly the BNCR rapidly became a Beyer Peacock railway.

It is impossible to overstate the Beyer influence on Irish locomotive development. For fifty years BNCR and NCC locomotive designs revealed their Beyer ancestry. The Great Southern and Western Railway's celebrated '101 class' were a pure Beyer design, and the similarity of the BNCR 'A class' to the GSWR's '60 class' is a further sign of common origins. On the Great Northern, Dundalk locomotive practice largely developed from the Beyer engines, while the Belfast and County Down Railway, the Sligo, Leitrim and Northern Counties Railway and the Cork, Bandon and South Coast Railway were almost exclusively

Beyer customers. Only two companies stayed outside the fold. The Midland Great Western – EL Ahrons' "most Irish railway" – had only six Beyer engines on its books, while the Waterford, Limerick and Western had none at all, preferring Dübs, Kitson and Vulcan. The narrow gauge was a different matter, with Beyer supplying only eight engines, all of them to the BNCR or their predecessors, the Ballymena and Larne. The engines built for the BNCR were, of course, the 2-4-2 compound tanks.

The BNCR and turf-burning

At some time in the early 1860s the BNCR experimented with turf as a fuel for locomotives. A train of seven carriages was run from Carrickfergus Junction (Greenisland today) to Limavady Junction, a distance of 74 miles. The combined weight of the train, including engine and tender, was 70 tons. With an engine weight of 25 tons, the motive power may have been either a a Sharp H class, or just possibly a BBC&PJR 2-2-2. Pressure at the start of the run was 100psi, rising first to 110psi and finally 120psi (the boiler's maximum pressure) on the climb from Carrickfergus Junction. The early account exaggerated the ruling gradient to 1 in 70; in fact 1 in 100 is the true figure. The fuel was compressed turf supplied by the Ballymena Compressed Peat Company. Figures claimed that turf consumption on this run was 21.47 lbs per mile, against an estimated average coal consumption, over a three month period, of around 25.25 lbs per mile. It seems to me that the turf figure is too low to be easily justified.

The train involved was probably the 9.30am from Belfast, and speeds of up to 40 mph were apparently achieved. Observing matters from a rather overcrowded footplate were the BNCR's Locomotive Superintendent Alexander Yorston, accompanied by Domville and Eaton, his counterparts on the Belfast and County Down and Ulster Railways respectively. From the figures alone, the observers concluded that peat was a very suitable locomotive fuel, but one suspects that only one side of the story was being told. A number of questions arise, one of which is why the experiment terminated at Limavady Junction on a train bound for Londonderry. Could it have been that, despite the very low consumption figures quoted, the engine ran out of turf before reaching its destination? It is surely significant that none of the three companies whose observers were present that day ever experimented further with turf as a fuel.

Table 1: Londonderry & Coleraine Railway locomotives

BNCR No	Builder	Built	Wheel arr	Driving wheels	Cylinders	Wdn	Notes
26	Fairbairn	1855	2-2-2WT	5'0"	13"x18"	1873	Ex-BBCPJR between 4/1858 & 1/1859
27	Longridge	July 1847	2-4-0	5'6"	15"x24"	1875	Ex-LER 2/1851 or 2/1852
28	SS 716*	3/1853	2-2-2WT	5'3"	11"x16"	1871	Sold to Mr Gregg, a contractor
29	SS 717*	3/1853	2-2-0WT	5'3"	11"x18"	1876	Rebuilt as 2-4-0 in 1869 with 14"x18" cylinders. Parts of No 32 used in the rebuild.
30	SS 722*	4/1853	2-2-0WT	5'3"	11"x18"	1880	
31	SS 723*	4/1853	2-2-0WT	5'3"	11"x18"	1878	
32	SS 718*	4/1853	2-2-0WT	5'3"	11"x18"	1869	Parts used in the rebuilding of No 29
33	Grendon	1/1859	0-4-2	5'0"	14"x20"	4/1889	Rebuilt 1871 with 16"x22" cyls
34	Fairbairn	1861	2-4-0	5'2"	14"x20"	11/1901	Rebuilt 1/1883

* The works numbers for specific 2-2-0WTs is uncertain.

Table 2: Ballymena, Ballymoney, Coleraine and Portrush Junction Railway locomotives

BBC&PJ No	BNCR No	Builder	Built	Wheel arr	Driving wheels	Cylinders	Wdn	Notes
1	22	SS 878	6/1855	2-2-2	5'6"	15"x20"	4/1906	Rebuilt as 2-4-0 in 1871
2	21	SS 879	6/1855	2-2-2	5'6"	15"x20"	1/1893	Rebuilt in 1870 (still 2-2-2)
3	22	SS 880	6/1855	2-2-2	5'6"	15"x20"	2/1877*	Rebuilt as 2-4-0 in 1871
4	23	SS 881	6/1855	2-2-2	5'6"	15"x20"	2/1886	Rebuilt as 2-4-0 in 1870
5?	24	Fairbairn	1855	2-2-2WT	5'0"	13"x18"	9/1898	Rebuilt in 1870
6?	25	Fairbairn	1855	2-2-2WT	5'0"	13"x18"	5/1883	Rebuilt as 2-4-0 in 1867
7?	-	Fairbairn	1855	2-2-2WT	5'0"	13"x18"	c1858	Sold to L&CR (see No 26 above)

* No 22 was withdrawn following the Moylena collision in 12/1876. Her boiler was sold.

Table 3: Early Belfast and Ballymena Railway locomotives

No	Name	Builder	Built	Wheel arr	Driving wheels	Cylinders	Wdn	Notes
1	Spitfire	SR 57	9/1839	2-2-2WT	6'0"	11"x18"	1863	Ex-UR 1847; renamed *Hawk* 12/1847; rebuilt as a 2-4-0T 10/1854
2	Vulcan	Bury	10/1847	0-4-2	5'0"	15"x22"	1869	Sold to Mr Gregg, a contractor
3	Gladiator	Bury	10/1847	2-2-2	5'6"	15"x20"	1868	Rebuilt as 2-4-0 in 4/1855
4	Hercules	Bury	10/1847	2-2-2	5'6"	15"x20"	1871	Rebuilt as 2-4-0 in 4/1854
5	Queen	Bury	10/1847	2-2-2	5'0"	15"x20"	1869	Rebuilt as 2-4-0 in 10/1853
6	Prince	Bury	10/1847	2-2-2	5'0"	15"x20"	1873	Rebuilt in 6/1862 (probably with a new boiler)
7	Hawk	SB 509	11/1847	2-2-2	5'0"	15"x20"	10/1849	To UR as *Spitfire* in exchange for No 1 (possibly in Dec 1847)
8	Falcon	SB 510	11?/1847	2-2-2	5'6"	15"x20"	1878	
9	Swallow	SB 511	11?/1847	2-2-2	5'6"	15"x20"	1887	
10	Kite	SB 512	12/1847	2-2-2	5'6"	15"x20"	1876	
11	Eagle	SB 513	1/1848	0-4-2	5'0"	16"x22"	1873	Boiler blew up 4/1857, reb 4/1858
7	Ostrich	SB 674	12/1850	0-4-2	4'6"	16"x24"	1872	Sold to Mr Gregg, a contractor

Note: The four Sharp 2-2-2s were Works Nos 509–12, but the order of names and numbers is uncertain.

Table 4: Later Belfast and Ballymena Railway locomotives

No	Builder	Built	Wheel arr	Driving wheels	Cylinders	Wdn	Notes
12	SS 936	4/1856	2-4-0	5'6"	15"x20"	5/1924	Rebuilt 1876, 1896, 1913
13	SS 937	5/1856	2-4-0	5'6"	15"x20"	11/1924	Rebuilt 1878, 1896, 1908
14	SS 938	5/1856	2-4-0	5'6"	15"x20"	11/1924	Rebuilt 1875, 1896, 1911; renumbered 14A 12/1922
15	SS 939	5/1856	2-4-0	5'6"	15"x20"	11/1924	Rebuilt 1875, 1890, 1910; renumbered 15A 12/1922
16	SS 940	6/1856	2-4-0	5'6"	15"x20"	1918	Rebuilt 1875, 1898; renumbered 16A? 4/1914
17	SS 941	6/1856	2-4-0	5'6"	15"x20"	9/1908	Rebuilt 1875, 1884; renumbered 17A? 1/1907
18	SS 993	5/1857	0-6-0	4'6"	16"x24"	9/1925	Rebuilt 1889, 1908
19	SS 994	6/1857	0-6-0	4'6"	16"x24"	10/1933	Rebuilt 1888, 1908, 1920 ('five foot' boiler)
35	SS 1277	7/1861	0-6-0	4'6"	16"x24"	11/1925	Rebuilt 1883, 1902 ('five foot' boiler)

Notes: Nos 12–17 rebuilt in 1875–78 with new boilers and frames, 15"x22" cylinders and thicker tyres.

Nos 12–17 rebuilt again in 1884–98, probably mostly with cabs. Possibly for new fireboxes in existing boilers.

Nos 12–15 rebuilt with 4' 0" boilers with flush fireboxes in 1908–13.

Table 5: Dimensions of early Belfast and Ballymena Railway locomotives

Class/numbers	Sharp No 1	Bury No 2	Bury Nos 3–6	Sharp Nos 7–10	Sharp No 11	Sharp No 7
Type	2-2-2WT	0-4-2	2-2-2	2-2-2	0-4-2	0-4-2
Cylinders	11"x18"	15"x22"	15"x20"	15"x20"	16"x22"	16"x24"
Coupled wheels	6' 0"	5' 0"	5' 6"	5' 6"	5' 0"	4' 6"
Leading wheels	4' 0"	–	4' 0"	3' 6"	–	–
Trailing wheels	4' 0"	3' 7"	3' 6"	3' 6"	3' 6"	3' 6"
Wheel base	6' 0" + 5' 6"	8' 0" + 6' 9"	7' 9" + 7' 4"	5' 9" + 7' 0"	7' 4" + 6' 3"	7' 9" + 6' 9"
Boiler length	8' 0"	11' 8"	11' 6"	10' 0"	10' 0"	11' 0"
diameter	3'6"	4' 0"	4' 0"	3' 7 ½"	3' 10"	4' 0"
tubes	149 x 1¾"	157 x 2⅛"	152 x 2⅛"	170 x 1¾"	170 x 1¾"	188 x 2"
Heating tubes	579 sq ft	1057 sq ft	1060 sq ft	798 sq ft		1102 sq ft
Surface firebox	50 sq ft	87 sq ft		79? sq ft		86 sq ft
Grate area	10.2 sq ft	15.2 sq ft	12¾ sq ft	12.6 sq ft	12¼ sq ft	14¾ sq ft
Boiler pressure	not known	80 psi	80 psi	80 psi	80 psi	100 psi
Tractive effort		6000 lbs	4600 lbs	5795 lbs	7012 lbs	9671 lbs
Weight		22½ tons	19½ tons	21½ tons		
Cost			£1912	£2010	£2160	£2160

Notes:

No 1 rebuilt as 2-4-0T with 5' 6" driving wheels and 3' 6" trailing wheels. Some accounts give 13" or 14" cylinders.

No 2 is recorded as having 16" cylinders, but ordered with 15" and they were 15" when withdrawn.

Nos 3/4 rebuilt as 2-4-0 with 5' 6" driving wheels and 16"x22" cylinders

No 5 rebuilt as 2-4-0 with 5' 0" driving wheels and 16"x22" cylinders

Table 6: Dimensions of later B & BR and BBC&PJR locomotives

Class/numbers	H class Sharp Nos 12–17	L class Sharp Nos 18, 19, 35	H class Sharp Nos 20–23	Fairbairn Nos 24–26
Type	2-4-0	0-6-0	2-2-2	2-2-2WT
Cylinders	15"x20"	16"x24"	15"x20"	13"x18"
Coupled wheels	5' 6"	4' 6"	5' 6"	5' 0"
Leading wheels	3' 6"	–	3' 6"	3' 6"
Trailing wheels	–	–	3' 6"	3' 6"
Wheel base	6' 0" + 7' 6"	7' 11" + 7' 4"	5' 9" + 7' 0"	c6' 6" + 6' 6"
Boiler length	10' 0"	11' 0"	10' 0"	c9'3"
diameter	3' 8"	4' 0"	3' 8"	
tubes	148 x 2"	187 x 2"	170 x 1¾"	
Heating surface	867 sq ft	1179 sq ft	868 sq ft	
Firebox casing	4'0½" x 4'5¾"	4'2¾" x 4'6"	3'7¾" x 4'8"	
Grate area	12.7 sq ft	14.3 sq ft	12¾ sq ft	
Boiler pressure	130 psi	120 psi	120 psi	
Tractive effort	6955 lbs	11605 lbs	6955 lbs	
Weight	c25 tons	38¾ tons (as reb)	21¾ tons	
Cost	£2460	£2560/£2490 (35)		£1888

Notes: Nos 12–17 rebuilt with new boilers and 15"x22" cylinders in 1875–78.

Nos 12–15 rebuilt with 4' 0" boilers in 1908–13, possibly second hand from K class 0-6-0s.

Nos 20, 22 and 23 rebuilt in 1870–71 as 2-4-0s with new boilers and 15"x22" cylinders.

No 21 rebuilt in 1870 with a new boiler but remaining as 2-2-2 with 15"x20" cylinders.

Nos 18, 19, 35 rebuilt with new boilers 4'3" x 11'0" and 17"x24" cylinders in 1883–89.

Nos 18, 19 reboilered in 1908 with 4' 0" boilers, 187 tubes, 14 sq ft grate, weight 38 tons 16 cwt.

No 35 given 4' 8½"x 10' 11" boiler in 1902, 193 tubes, 14½ sq ft grate, BP 170psi, 4' 7½" driving wheels, TE 18,058 lbs

Table 7: Dimensions of Londonderry and Coleraine Railway locomotives

Class/numbers	Longridge No 27	Sharp Nos 28–32	Grendon No 33	Fairbairn No 34
Type	2-4-0	2-2-0WT	0-4-2	2-4-0
Cylinders	15"x24"	11"x18"	14"x20"	14"x20"
Coupled wheels	5' 6"	5' 0" (later 5' 2")	5' 0"	5' 2"
Leading wheels	3' 9"	3' 0"	–	
Trailing wheels	–	–		–
Wheel base	5' 2½" + 5' 11"	10' 0"		
Boiler length	13' 7"	9'6"		
diameter	3'9" (HS 948 sq ft)	3'2" (HS 573 sq ft)		
tubes	123 x 2"	111 x 1⅞"		
Grate area	10½ sq ft	10½ sq ft		
Boiler pressure	90 psi		120 psi	120 psi
Tractive effort	6259 lbs		6664 lbs	
Weight	23 tons, 12 cwt	12 tons		

Notes: No 29 rebuilt as a 2-4-0 with 14"x18" cylinders and 5'3" driving wheels in 1869, using parts from No 32.

No 33 was rebuilt with 16"x22" cylinders in 1871

Chapter 2
The Early Years of the BNCR: 1861–1876

Like most companies formed by amalgamations of smaller concerns, the BNCR made no drastic changes at first. Its first two engines were L1 class 0-6-0s numbered 36 and 37, built by Beyer Peacock (makers numbers 365 and 366) in 1863 at a cost of £2400 each. No 36 lasted until 1932, though No 37 was withdrawn in 1928. The major dimensions of these engines appear in Table 9. It is rather an oddity that these engines were classified as L1, since they actually bore more similarity to the K class. A brief discussion of the classification system of BNCR, and later NCC, engines appears at the end of this chapter.

Along with their earlier L class cousins, these engines marked the beginning of the six-coupled era on the BNCR, and the supersession of the 0-4-2 as the standard goods engine type. Only the GSWR and MGWR preceded the BNCR in the development of the 0-6-0 type, and the L and L1 classes continued on main line work until 1923.

Continued on page 25

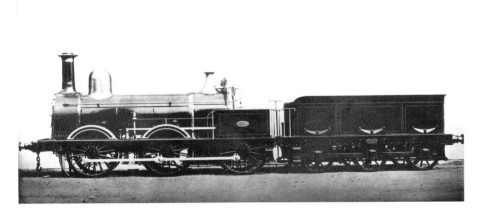

The BNCR shopped around for its next two goods engines and this time Beyer Peacock got the order. The Beyer engines were similar to the L's, but with 5'1⅝" wheels – a size which (with variations due to tyre thickness) became a BNCR/NCC standard for 0-6-0s. The photograph is the Works picture of No 36. Note the absence of a cab or coupling rod splashers and how the centre of the splashers is open. No 36 was the first Beyer engine on the BNCR.

Beyer, Peacock

A much later view of No 36 piloting a 2-4-0 on a goods near Whiteabbey in the 1890s. Note the length of the chimney, the raised firebox and the feed water pipe beautifully curved over the splasher.

LGRP

Left: The L1 class engines were reboilered in 1877/79 and received larger cylinders in 1895/94 respectively. In 1904/05 they were again rebuilt, this time with 5'0" boilers. This official view shows No 37 at Belfast either on rebuilding in 1905 or at its first general overhaul after that event. Amazingly, the splashers are still open.

Official NCC photograph

Right: Sister engine No 36 is receiving attention at Belfast in July 1932. Under the Midland most NCC engines had received extended cabs.

JAGH Coltas

Below: A final view of No 36 working on a ballast train in connection with the new loop line, near Whiteabbey in September 1932. She was withdrawn three months later.

RS Holden

(Continued from page 23)

With the rapid increase in goods traffic, two further 0-6-0s were tendered for in 1867 and this time Sharp Stewart got the order. They were numbered 38 and 39 and cost £2390 each, arriving in June 1867. The K class 0-6-0s eventually numbered nine engines – a large class for the nineteenth century BNCR.

The first five had 16"x24" cylinders, as built, but the last four had 17" cylinders from new. These four also had cabs and Nos 43 and 44 had Duplex safety valves. The final two engines, Nos 31 and 30, were built by Beyer Peacock. These last two had 4'1" diameter boilers, all the earlier engines having 4'0" boilers, and were easily distinguished by the double beading on the splashers, which once carried the Beyer, Peacock builder's details (see page 27). The details and number sequence of this class can be found in Table 8, and the dimensions in Table 9.

Table 8 shows that after the construction of the first two Ks, Nos 38 and 39, the BNCR broke from a strict numerical sequence and began to use numbers left vacant by the scrapping of older engines. Thus, Nos 32 (1870) and 28 (1871) replaced ex-L&CR 2-2-2WTs, and No 7 (1873) replaced an ex-B&BR 0-4-2. Nos 43 and 44 (1875–76) had new numbers and, finally, the Beyer engines, Nos 31 (1878) and 30 (1880), replaced the last two ex-L&CR 2-2-2WTs. Thus began the

Continued on page 27

The only surviving view of a K class 0-6-0 in original condition is this one of No 28 prior to rebuilding in 1898. No 28 was a Sharp engine of 1871. Sand boxes are placed in a convenient position. The dome is brass and there is a massive brass ring around the raised firebox. A bent weather board provides minimal comfort for the crew.

Author's collection

Representing the class in the second stage of their lives is No 32 of 1870 after rebuilding in 1893, with short cab and flush firebox boiler, but still retaining a sloping smoke box. The immaculate turn out of engine and the posed view, with two gentlemen on the ground wearing hats and the crew on the footplate, suggest that something special is being recorded. Note the sandboxes below the footplating.

Locomotive Publishing Co

Above: No 7, after rebuilding in 1896, still has the sand boxes above the footplating and has a tablet exchanger on the tender. The sloping smoke box is still there, as are a standard whistle and a crowing whistle.

Locomotive Publishing Co

Above: Most of the K class were transformed from 1909 onwards by the fitting of 5'0" boilers. This shot of No 32 was taken in 1917, when it was rebuilt. Note the extended cab roof with hand rails, Ross 'pop' safety valves, tender coal rails and rear toolbox. *Author's collection*

Right: The first two members of the K class were Nos 38 and 39, built in 1867. They received five foot boilers in 1912 and, as a further development, No 38 was fitted with a Phoenix superheater, requiring an extended smokebox. Both were soon removed.

Official NCC photograph

Above: Dating from 1873, K class No 7 was rebuilt as a K1 class in March 1923 sporting a unique cab with deep roof valances, quite unlike anything else on the NCC. It is often claimed that the NCC '7' was an inverted '2' but close study of the NCC '2' shows that this is not the case. The '7' has no straight lines on it at all, whereas the '2' (as seen in the picture of No 28) has a straight bottom bar and is more slender.

Author's collection

Left: This view of No 7 at Londonderry, on 14 July 1931, shows another unique feature of No 7 – a circular handle on the smokebox door similar to the Glasgow and South Western Baltic tanks. Note the typical small BNCR brake van behind the engine.

Real Photographs Ltd

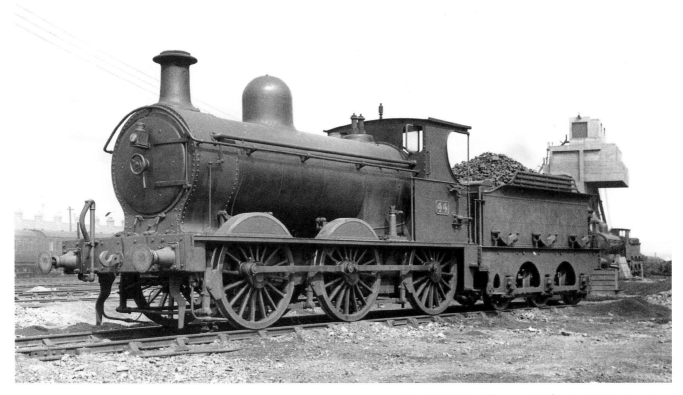

Above: K1 class 0-6-0 No 44, at Belfast on 7 August 1936, illustrates the final condition of most of the class. In the background is the new concrete coaling tower, only recently installed.

RG Jarvis,
Midland Railway Trust, Ltd

Left: The Beyer pair had unique splasher beading, as seen on No 31 in Coleraine yard on 22 June 1937. Note the small 'NCC' lettering and the smokebox door handrail.

HC Casserley,
courtesy RM Casserley

(Continued from page 25)

chaotic numbering system which was to characterise the BNCR and its NCC successors! In fairness, it must be added that most contemporary railways took similar steps to fill up gaps in their numbering sequences as they arose. The really odd thing, of course, was that the number 1 was left vacant from 1863 until 1869.

As with the earlier 0-6-0s, the K class went through several stages of rebuilding, as outlined in Table 8 and in the captions. Boilers with flush-topped fireboxes replaced the original raised-firebox type, all received cabs and in their final condition all, except No 28, had the 'five foot' boiler. The early versions of this had a small 14½ sq ft grate but, from 1920 onwards, a larger 18¼ sq

ft grate was fitted. This version was classified as 'K1'.

Pari passu with the new goods engines came I class 2-4-0s Nos 40 and 41 in 1868. These, the first new passenger locomotives for the BNCR, cost £2050 each and were described by their builders, Beyer Peacock, as the '2268 class', (after their order number). Numbering complexity continued with this class. In 1869, they were renumbered 1 and 2, as the earlier bearers of these number were scrapped. No 1 was reboilered and fitted with vacuum brakes in 1892. She was rebuilt again in 1911 with a new firebox in her existing boiler, lasting in this form until September 1924. No 2 was reboilered in 1891. She was rebuilt again with a new firebox in

I class 2-4-0 No 1, as delivered by Beyer Peacock in 1868. Maker's plates are on the tender frame and around the splasher. Crew comfort is minimal with only a weatherboard. The wheels are 5'7", so to the same scale the chimney must be at least four feet tall. The engine has a similar appearance to larger 2-4-0s supplied to the Dublin and Belfast Junction railway in 1866. Beyer Peacock at this time were turning out a standard 2-4-0 called the '838 class' of which the Dutch state railways bought 74. The same general design with different wheel sizes went to Ireland.

LGRP

Above: I class No 2 is seen here after first rebuilding in 1891. Obvious changes from the previous picture are the cab, Ramsbottom safely valves in place of Salter, the fitting of a vacuum brake and the tablet enchange apparatus mounted on the tender. Note the large chest perched on the tender.

Author's collection

Right: I class No 2 shunting at Magherafelt on 4 May 1920. Coal rails have been added to the tender and the Beyer lettering has been removed from the splasher. The signalman takes the chance to get into the photograph while the engine blows off lightly.

Ken Nunn Collection, LCGB

her existing boiler and fitted with the vacuum brake in 1910, being finally withdrawn in December 1924. The dimensions of this class appear in Table 10.

The I class were little better than the Sharp Stewart 2-4-0s of twelve years previously but, as already mentioned, the BNCR was in desperate straits for motive power in 1868. Further evidence of panic in the Locomotive Department came in 1870 when York Road decided to build its very own first engine. This was a modest beginning – a little 2-4-0 with 15"x20" cylinders, 5'6" diameter driving wheels and 3'6" leading wheels. It took the number, and probably also the cylinders, of the old Bury 2-4-0 No 5 (originally a 2-2-2) which was scrapped in 1869. It is doubtful whether York Road at this time could have cast cylinders, and using the existing ones would have cut down costs. Pattern making for a cylinder block was a highly skilled job

and an iron foundry capable of producing a casting for a cylinder block, weighing over a ton, would have been a major investment. No 5 was reboiled in 1887 and withdrawn in 1907.

Flushed with success, the shops turned out another similar 2-4-0, once again using the number and the cylinders of scrapped Bury 2-4-0 No 4. The similarity of the York Road-built engines to the Beyer Peacock I class suggests that York Road regarded them as part of that class. The new engines cost £1504 2s 6d each. (I was amazed at the precision of the accounting until the late Harold Houston pointed out that the BNCR was a private company which had to pay dividends to its shareholders, and not a nationalised railway which could squander its resources!)

Like the original Bury engine, No 4 had 4'1¼" leading wheels, probably reused, and was the only I class engine

No 4 at Londonderry about 1898. She was built at York Road using cylinders and leading wheels from a scrapped Bury 2-2-2 engine. No 5 was similar but had smaller leading wheels. Both were regarded as I class. No 4 has been fitted with a vacuum brake and a tender weather board, probably for the Dungiven branch. The open splasher braced in the middle is unusual. Bob Clements commented that the cab and number plate, with company initials and building date, are exactly to the GS&WR Inchicore pattern. It is possible that Edward Leigh, who succeeded Yorston from 1868 to 1875, used drawings from Inchicore for this rebuild. The cab pattern was also used for K class 0-6-0 Nos 43 and 44, built in 1875/76.

Locomotive Publishing Co

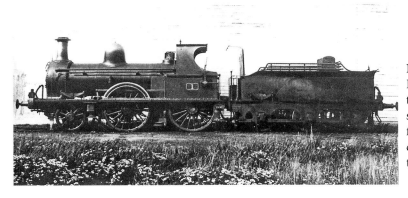

No 1, photographed at the traditional spot in Portrush, also has a tender weather board and a tablet catcher mounted on the cab. Ramsbottom safety valves have been fitted. This view is after final rebuilding in 1911 and the tender now has coal rails fitted. Note how the reversing rod runs through the driving wheel splasher.

Official NCC photograph

On 6 May 1920, No1 is seen at Limavady Junction with the branch train to Dungiven and showing the correct lamp code. She is in final condition with steam heating equipment fitted. The train of three six-wheelers will not tax even a 60 year old 2-4-0 on the heavily graded Dungiven branch. The fine display of chimney pots on the station building is worth a look.

Ken Nunn Collection, LCGB

to have leading splashers. She was reboiled in 1888, fitted with the vacuum brake in 1891, renumbered 4A in 1903 (an unusual example of the BNCR using a duplicate number) and withdrawn in May 1905. In her latter days she could often be found on the Dungiven branch. (By coincidence, a much later No 4, of class U1, was renumbered 4A when WT 2-6-4T No 4 appeared. The 'A' was stencilled on the front buffer beam and above the number plates.)

Since the BNCR was using bits from the old Bury locomotives in Nos 4 and 5, it is possible that Bury No 3, which went to Grendons to be rebuilt, and of which no photographs exist, was also similar in style to the I class.

York Road's third excursion into locomotive building occurred in 1873 when M class 0-4-2 tender engine No 26 was turned out (the number of a Fairbairn 2-2-2WT scrapped in 1873). The large trailing wheels

look identical to the leading 12 spoke wheels on I class 2-4-0 No 4, and could well have come from a scrapped Bury engine. Few, if any, parts of the original No 26 can have been used, so this was probably York Road's first totally new engine. She was rebuilt in 1892 (new boiler and cylinders), 1903 ('five foot' boiler) and 1908 (17"x24" cylinders). She was the last engine to be turned out in Belfast for almost twenty years, and her dimensions can be found in Table 12.

It is a mystery why an 0-4-2 was built as late as 1873, when York Road had been using 0-6-0s for 16 years, but it is possible that she may have been intended as a banking engine for heavy trains on the Greenisland to Ballyclare Junction section, in the days when main line trains reversed direction at Greenisland. In her latter years as the Ballymena pilot, No 26 was often used to bank trains up to Galgorm and on to Cullybackey. She was withdrawn in 1925.

This photograph of No 26 dates from the late Midland era, after 1918, when engines had 'M R' on the tender flanking 'NCC' in scroll lettering. The engine was fitted with a 'five foot' boiler in 1903. Note the sand box located under the footplating just behind the front buffer beam. NCC pattern balance weights can just be seen on each wheel below the splashers.

Official NCC photograph

Of much more importance than the three home-built products were the nine G class 2-4-0s, turned out between 1872 and 1878 to see the BNCR through a particularly difficult period. The last two, Nos 8 and 22, were built by Beyer Peacock, and the first seven by Sharp Stewart. Beyers referred to their progeny as 'class 8', after the first of the two engines they built. Further details appear in Tables 8 and 10.

The G class continued the complexity of the BNCR's numbering scheme. Nos 6, 8 and 10 took the numbers of scrapped B&BR 2-2-2s, No 11 took the number of a B&BR 0-4-2, No 22 was the original number of a BBC&PJR 2-2-2, No 27 of the L&CR Longridge 2-4-0, and No 29 of a L&CR 2-4-0 rebuilt from a 2-2-0WT. The first two engines of the class, No 40 and 41, used vacant numbers, though these had been briefly used in 1868

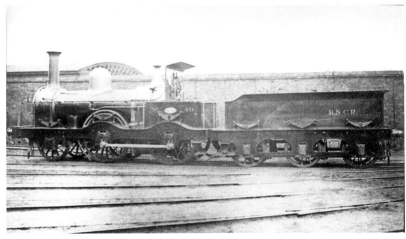

Left: This shot, probably dating from the 1870s, shows G class No 40 as delivered from Sharp Stewart. The weather board offers minimal protection. The shot reveals detail of the BNCR livery of this era. The number is painted on – no plate as yet and 'BNCR' is painted on the rear of the tender.

CP Friel collection

Below: After reboilering in 1897, No 40 has now got her numberplate and a small cab. Ramsbottom safety valves and a tablet catcher mounted on the tender have been fitted. A footstep is attached to the heavy front axle box. Note the absence of smokebox wing plates.

Author's collection

Opposite left: M class 0-4-2 No 26 was built at Belfast in 1873 but no picture of her in original condition has survived. This photograph was probably taken in 1908 when she was rebuilt with 17"x24" cylinders. She has a beautifully polished brass dome and is in early Midland era livery with 'NCC' in small letters on the tender and the Midland crest on the cab side. The fireman, leaning on the hand brake has a handsome gold chain on his waistcoat, but possibly no watch in the pocket. Few fireman could afford a watch around 1905.

Official NCC photograph

G class Nos 8 and 22 were built by Beyer Peacock in 1878. No 8 is seen here at Londonderry shed, probably shortly after first rebuilding in 1896. The engine is painted in 'invisible green' with a brass dome. The Beyer Peacock plate can be seen on the frames below the leading splasher. The mass of rivets would challenge a modeller and suggest that the two Beyer engines had sandwich frames.

LGRP

An unidentified G class on a fairly heavily loaded passenger train, north of Ballymena around 1900. The train is clearly in two portions, perhaps a mainline bogie set heading for Londonderry, combined with a Portrush-bound six-wheel set at the rear.

LGRP

This view shows G class No 29 after rebuilding in 1899 when it was fitted with a flush firebox casing. The front step is mounted behind the leading wheel axle box.

Author's collection

G class Nos 6, 10 and 27 were rebuilt with 'five foot' boilers as Class G1 in 1910–13. Here, No 27 shunts a six-wheel coach onto her train at Portrush on 11 July 1931. The white diamond below the chimney indicates a branch line train. Note the smokebox wingplates.

Real Photographs Ltd

An official view of newly rebuilt G1 class No 6 at Belfast in Nov 1913. Ross pop valves have been fitted and, in contrast to the Beyer engines, the Sharp examples have plate frames. Once again, the engine is immaculate. R Welsh normally received the commissions for official photographs. *Official NCC photograph*

Left: G class No 41 at Londonderry on 6 August 1930 showing the final condition of those engines which had small boilers. She has smokebox wingplates and has retained Ramsbottom valves but the brake blocks are almost done. Note the small 'LMS NCC' lettering on the tender, but otherwise the engine is still in MR(NCC) livery. The white-painted rig behind the engine is some sort of hoist.

HC Casserley, courtesy RM Casserley

for the first two members of the I class (see page 27). The G class were long-lived and extremely useful little engines. According to the late Harold Houston's father, No 27 was a familiar sight on the Portrush branch until as late as 1933, while Nos 6, 10 and 11 could be found on the Larne line in the early 1920s and No 41 had a spell at Dungiven during the First World War.

One locomotive remains to be mentioned from the BNCR's first 18 years as a company. This was N class No 42, a 0-4-0ST dock shunter (Sharp Stewart 2444 of 1875), which cost the Company £2150. The Belfast Central Railway had opened in 1875 and, although it carried little passenger traffic, it formed an important link for freight traffic between the Belfast docks and the termini of the BCDR, the BNCR and the GNRI. The tightly curved harbour lines beside the steamer terminal sheds, and the cramped tunnel beside the Queens Bridge, required a powerful 0-4-0ST with a short wheelbase. Her driving wheels were 4'0⅛" diameter, and her wheelbase 7'9". Cylinders were 16"x22", and boiler 11'6" long and 3'6½" in diameter. There were 118 tubes of 1⅞" diameter, and the engine's heating surface was 661 sq ft. Tractive effort was 12,965lbs, and the engine weighed 30 tons.

The life of a shunting engine differed from its more glamorous main line sisters. For a start, conventional mileage records were meaningless. For No 42 and the later 'shunting pugs', mileage was calculated on the

The opening of the Belfast Central Railway in 1875 created the need for a shunting engine able to negotiate the tight curves around Belfast docks. Sharp Stewart supplied N class 0-4-0ST No 42 in 1875. She is in original condition in this picture, without cab. For her size, she was powerful and, at 31 tons, had plenty of adhesion. Note the Sharp Stewart works plate on the saddle tank.

Author's collection

basis of a 5mph average, and overhauls were carried out every three years. On this basis, No 42 ran about 900,000 miles during a busy life. Only a handbrake was fitted to the engine, and the fireman was expected to work it by spinning a large wheel in the cab. To save themselves a lot of work, most men left the wheel set in a position where the brakes were just rubbing. They had to be careful, though, as some irascible old drivers on spotting this would immediately spin the wheel to the full 'off' position, thus creating more work for their long-suffering mates.

To avoid buffer-locking on the sharp dock curves, No 42 had rectangular buffers with the corners rounded off. In 1901 her crews were given the luxury of a cab

– but as this enclosed the safety valves it was extremely noisy! It was perhaps as well that they stayed alert, for the Belfast Harbour Commissioners produced all sorts of restrictions on the tramway section. Engines were not permitted to blow off, the drain cocks could not be opened and, since the tramway crossed a public road, engines had to be preceded by a man with a red flag (and an additional flagman to look after the train if the load exceeded two wagons). Finally, the engine was forbidden "to be nearer than 15 yards to any other train approaching or preceding on the same line of rails." What evasive tactics could be employed is not suggested!

This later picture, though still in BNCR days, shows her with a cab and several more dinges in her cylinder covers, indicative of a hard life down the docks. The lamp code shows 'empty carriage train' so No 42 may have been on station pilot duty. Jennymount Mill dominates the background.

Author's collection

The BNCR locomotive position in 1876

The choice of 1876 to end the first period of BNCR locomotive history is not entirely arbitrary, for in this year the Company appointed as its new Locomotive Engineer Bowman Malcolm, who was to hold the post for the next 47 years. Before considering his massive achievements, it is worth considering what Malcolm inherited. There were 23 2-4-0s (with two more G class to come in 1878), 12 0-6-0s (with two others to come in 1878 and 1880), two 0-4-2s (Nos 26 and 33), three 2-2-2s (Nos 8 and 9 from the B&BR and No 21 from the BBC&PJR), three 2-2-2WT (No 24 from the BBC&PJR and Nos 30 and 31 from the L&CR) and one 0-4-0ST, No 42 the dock shunter just described.

The BNCR was, therefore, not badly placed for locomotives. Of the passenger engines, the H class had all been rebuilt in the mid 1870s, while the G class were virtually new. The I class and the Grendon 2-4-0 were less than ten years old, and only the single drivers could be considered as life-expired. Of the goods engines, the K class were all less than ten years old, while the L class, though older, were excellent engines. The younger L1 class were just coming up for reboilering. With the average age of his engine stock at well under 20 years, Malcolm could take time to play himself in gradually.

Since 1876 can be considered a watershed for the BNCR, comparison with the other leading Irish railways is instructive, beginning with the Great Northern Railway which was formed by amalgamation in the same year. Like the BNCR, the GNRI's standard passenger type was the 2-4-0, though it still had a large stock of 2-2-2s. For goods it had a few 0-6-0s, but relied heavily on the 0-4-2 type, many of which lasted until the early 1900s when the GNRI's standardisation programme replaced them.

On the southern lines, the Great Southern and Western Railway was still building 2-4-0s for its passenger trains. By 1876 about half of what became a 118-strong '101 class' had been built, so it was well equipped with 0-6-0s for freight duties. The neighbouring Midland Great Western Railway was also a 2-4-0 railway for passenger trains, but relied on 0-4-2s for goods traffic; its first standard 0-6-0 appeared in 1876. Both the cross-country Waterford, Limerick and Western Railway and the Dublin, Wicklow and Wexford Railway also employed a preponderance of 2-4-0s and 0-4-2s, but the latter's more suburban status was reflected in the higher number of tank engines it owned.

Thus, all the Irish railways in 1876 relied on 2-4-0s for passenger work, but opinions were divided between 0-4-2s and 0-6-0s for goods. The BNCR was firmly in the latter camp, with 85% of its goods engines being 0-6-0s. Yet, paradoxically, apart from two more 0-6-0s built in 1892, it was to build no more of this type between 1880 and 1923.

The demise of the single driver was inevitable on most Irish railways. By 1900, coupled driving wheels signified progress, but more important, reflected the increasing weight of trains to be worked. The last railway to use single drivers on important passenger trains must have been the Great Northern. Their very fine 4-2-2s Nos 88 *Victoria* and 89 *Albert* remained on main line duties until 1905 – not due to lack of progress, but rather to the well laid, and moderately graded, main line south of Dundalk.

At the end of this chapter, a word might be said about the classification letters which were to identify NCC locomotives from the early twentieth century onwards. The introduction of this system may, or may not, have been the idea of the Midland Railway when they took over the BNCR – though if it was, one wonders why they never applied it to their own fleet of 4' 8½" gauge engines. My own guess is that as engine numbers increased and classes diversified, someone (possibly a junior in the locomotive department) was given the task, around 1902, of devising a logical system. My conjecture is that he would have designated the newest passenger engine – a 4-4-0 compound – **A**, then **B** for a smaller engine of a similar type, **C** and **D** for earlier compounds with six foot and seven foot wheels respectively. This left the 0-6-0 compounds, Nos 53 and 54, to be classified **E**. Older engines then followed, with the earlier 2-4-0s classified from **F** to **I** and the 2-4-0 tanks **J**, followed by the goods engines, **K** and **L**, following which some mavericks appeared – **M** for the 0-4-2 and **N** for the dock shunter! Letters **0** to **S** were allocated to narrow gauge engines, thus completing the system. Subsequently the 1905 railmotors were **T** (reused in 1924 for the Ballycastle 4-4-2Ts), the 1914 simple 4-4-0s **U** and the 1923 0-6-0s **V**. The LMS continued the system, using sub-classes like **A1**, **B3** and **U2** for rebuilds and variations. They used **W** and **WT** for the moguls and 'jeeps', the 1937 diesel shunter was **X** and the regauged 'Jinties' **Y**. Finally, the UTA used **Z** for the two SLNCR engines purchased in 1959 after the closure of that company.

During the period covered by this chapter, though, the problem was that locomotive builders identified engines they supplied by the number of the first locomotive in the class. Hence what became the G class appears in Beyer Peacock records as class 8, and the F class engines as the 45s.

Table 8: Chronology of L1, K, I, G, M and N classes, built 1863–80

Class	Type	No	Deliv	Builder	Cost	Rebuilt (1)	Rebuilt (2)	Rebuilt (3)	Rebuilt (4)	Scrapped
L1	0-6-0	36	7/1863	BP 365	£2400	10/1877	1895 (17")	2/1904 (LB)	1917	12/1932
L1	0-6-0	37	8/1863	BP 366	£2400	6/1879	1894 (17")	4/1905 (LB)	1914	11/1928
K	0-6-0	38	6/1867	SS 1797	£2390	2/1884 (17")	1896	8/1912 (LB)	12/1930 (K1)*	6/1938
K	0-6-0	39	6/1867	SS 1798	£2390	7/1878 (17")	1898	4/1912 (LB)	2/1924 (K1)	6/1927
K	0-6-0	32	6/1870	SS 2001	£2390	? (17")	1893	1905 (LB)	1917	7/1933
K	0-6-0	28	6/1871	SS 2143	£2390	2/1884 (17")	1898	1910		9/1925
K	0-6-0	7	6/1873	SS 2269	£2288	12/1882 (17")	1896	1909	3/1923 (K1)	10/1934
K	0-6-0	43	9/1875	SS 2487	£2260		1894	1906	1920 (K1)	6/1938
K	0-6-0	44	4/1876	SS 2630	£2730		12/1892	3/1909 (LB)	6/1921 (K1)	12/1929*
K	0-6-0	31	6/1878	BP 1712	£2510		1896	1907	1927 (K1)	8/1947
K	0-6-0	30	6/1880	BP 1920	£2510		1897	1909	8/1922 (K1)	6/1938
I	2-4-0	1	11/1868	BP 850	£2050		1/1892	1911		9/1924
I	2-4-0	2	11/1868	BP 851	£2050		8/1891	1910		12/1924
I	2-4-0	3	9/1868	Grendon						12/1902
I	2-4-0	4	1871	YR		5/1888				5/1905
I	2-4-0	5	12/1870	YR	£1504	11/1887				1/1907
G	2-4-0	40	6/1872	SS 2218	£2238		1897	1911	1920	12/1925
G	2-4-0	41	6/1872	SS 2219	£2238		1895	1910	1923	7/1933
G	2-4-0	6	12/1873	SS 2335	£2830	1887	1896	1913 (G1)		12/1931
G	2-4-0	11	12/1873	SS 2336	£2830		1895	1904	1921	9/1933
G	2-4-0	10	4/1876	SS 2627	£2620		1894	3/1909 (G1)		12/1931
G	2-4-0	27	4/1876	SS 2628	£2620		1894	10/1910 (G1)	1925	7/1933
G	2-4-0	29	4/1876	SS 2629	£2620		1895	1908		11/1925
G	2-4-0	8	6/1878	BP 1713	£2830		1896	1906		12/1930
G	2-4-0	22	6/1878	BP 1714	£2830		1896	1904		11/1928
M	0-4-2	26	6/1873	YR			5/1892	1903 (LB)	1908 (17" cyls)	6/1925
N	0-4-2ST	42	1/1875	SS 2444				1901		6/1925

Notes: LB indicates 'large boiler' – ie locomotives fitted with the 4' 8¼" (or 'five foot') boiler. Classes K1 and G1 also had this boiler.

K class: 1878–84 – first five engines given 17" cylinders

1892–98 – all, except 38/39, given new cylinders, cabs and flush-topped 4' 0" boilers

1905–12 – Nos 32, 38, 39 and 44 given 4' 8¼" boilers, and No 38 fitted with a Phoenix superheater

1920–27 – Six engines given 4' 8¼" boilers with 18¼ sq ft grates and reclassified K1

* 1930 – Nos 38 and 44 combined using the frames and motion of No 38, and the K1 boiler from No 44.

Table 9: Dimensions of L1, K and K1 classes

Class/numbers	L1 class	L1 class with 4' 8¼" boiler	K class built 1867–73	K class built 1875–80	K class final condition*	K1 class
Type	0-6-0	0-6-0	0-6-0	0-6-0	0-6-0	0-6-0
Cylinders	16"x24"	17"x24"	16"x24"	17"x24"	17"x24"	17"x24"
Coupled wheels	5' 1⅝"	5' 2⅝"	5' 1⅞"	5' 1⅞"	5' 2⅝"	5' 2⅝"
Wheel base	7' 3" + 7' 5"	7' 3" + 7' 5"	7' 0" + 7' 10"	7' 0" + 7' 10"	7' 0" + 7' 10"	7' 0" + 7' 10"
Boiler length	11' 8"	10' 11"	11' 0"	11' 0"	11' 0"	9' 10"
diameter	4' 0"	4' 8¼"	4' 0"	4' 0"/4'1"	4' 0"	4' 8¼"
tubes	187 x 2"	193 x 2"	187 x 2"	156/152 x 2"	187 x 2"	193 x 2"
Heating surface	1188 sq ft	1157 sq ft	1180 sq ft		1180 sq ft	
Firebox casing	4' 2½" x 4'6"	4'4" x 4'5½"	4' 2¾" x 4'5½"	4' 2¾" x 4'5½"		
Grate area	14½ sq ft	14½ sq ft	14½ sq ft	14½ sq ft	14 sq ft	18¼ sq ft
Boiler pressure	120 psi	170 psi	130 psi	130 psi	170 psi	170 psi
Tractive effort	10,273 lbs	16,004 lbs			16,004 lbs	16,004 lbs
Weight	27 tons, 15 cwt	38¾ tons	33 tons, 4 cwt	33 tons, 4 cwt	35 tons	39 tons, 14 cwt

Notes: Class L1 – TE 16,297 with 17"x24" cylinders after 1894/5 rebuilding

Class K – Nos 8 and 22 were built with 4' 1" boilers

*Class K – Nos 38, 39 and 44, with 4' 8¼" boilers but only 14½ sq ft grate, weighed 38 tons, 9 cwt

Table 10: Dimensions of I, G, G1, M and N classes

Class/numbers	I class	G class as built	G class final condition	G1 class	M class	N class
Type	2-4-0	2-4-0	2-4-0	2-4-0	0-4-2	0-4-0ST
Cylinders	15"x20"	16"x22"	16"x22"	16"x22"	16"x24"	16"x22"
Coupled wheels	5' 6"	5' 6¼"	5' 7½"	5' 7½"	5' 1½"	4' 0"
Leading wheels	3' 6½"*	4'1½"	4'1½"	4'1½"	–	–
Trailing wheels					4'1"	–
Wheel base	6' 6" + 7' 6"	6' 9" + 7' 7½"	6' 9" + 7' 7½"	6' 9" + 7' 7½"	8' 0" + 7' 7"	7' 9"
Boiler length	9' 10"	10' 0"	10' 0"	9' 11"	11' 0"	11' 6"
diameter	3' 9"	3' 9"	3' 8"	4' 8¼"	4' 0"	3' 6"
tubes	164 x 2"	125 x 2"	125 x 2"	193 x 2"	187 x 2"	122 x 2"
Heating surface	947 sq ft	749 sq ft	794 sq ft	1073 sq ft	1188 sq ft	821 sq ft
Firebox casing	4'1" x 4'7"	4'4" x4'6½"		4'4" x4'5½"	4'4" long	3'6" long
Grate area	14 sq ft	14½ sq ft	14½ sq ft	14½ sq ft	14½ sq ft	11½ sq ft
Boiler pressure		130psi	150psi	150psi	140 psi	130 psi
Tractive effort			10,638 lbs	10,638 lbs	11,675 lbs	
Weight	25½ tons	32 tons, 8 cwt	34 tons, 9 cwt	38 tons, 1 cwt	31½ tons	29 tons, 13 cwt

Notes: Class I after 1910–11 rebuilding – 134 tubes, BP 140lbs, grate 13¼ sq ft, weight 29 tons 4 cwt

* Class I Nos 4/5 – These engines had second hand leading wheels; No 4 – 4' 1¼", No 5 – 3'6"

Class M in final condition – Cyls 17"x24", 4' 8¼" x 9'10" boiler, TE 16,004 lbs, weight 39 tons, 8 cwt

Class N in final condition – BP 130 psi, TE 12,707 lbs, weight 31 tons

Chapter 3
Bowman Malcolm – The Early Days

Bowman Malcolm joined the BNCR in 1870 at the age of sixteen. He had no formal engineering qualifications but made such a good impression in his six years in the Locomotive Department that he was appointed Locomotive Superintendent in 1876, in place of Robert Finlay who had replaced Leigh in 1875. Only Daniel Gooch on the Great Western Railway had achieved a similar promotion at an earlier age (in his case aged 21), though his twenty-seven year reign at Swindon was as nothing compared to Malcolm's forty six years at York Road.

To be appointed Locomotive Superintendent of the fifth largest railway in Ireland at such an early age was quite remarkable, and indicates his ability as an engineer. He was certainly lucky to be in the right place at the right time, and the speed of his promotion may well have been due to his connections with Edward John Cotton, the long-serving Company chairman. Another story suggests that his predecessor Robert Finlay, who resigned due to ill health, had in fact died in office, and that Malcolm's appointment was initially a temporary one which was soon confirmed as permanent. Before considering his early designs, a flavour may be given of the personality who dominated the BNCR for so long.

Malcolm was a strict, though fair, disciplinarian by the standards of the Victorian age. Billy Hanley, who began his NCC career as a fourteen year old cleaner in 1911 and finished as Chief Locomotive Inspector in 1962, encountered Malcolm frequently in his early days. He was passed out for firing in 1916. Shortly afterwards he encountered Malcolm at Ballymena shed, and the great man subjected him to close questioning on the difference between simple and compound expansion. Evidently young Hanley impressed his chief, because some weeks later he attempted to join the army with a group of his mates. As he approached the Recruiting Office he felt a tap on the shoulder and heard a familiar voice: "Not you – you are more use to me on the railway!" Malcolm realised the worth of one of his best men. Several of those who joined up that day never returned from the Western Front, and in subsequent years Hanley was to say: "He probably saved my life."

Not that his own promotion was at anything like the speed of his boss, for Hanley had to wait until 1933 before he was passed for driving. But long experience enriched his future career, and throughout his life, even in his years as an Inspector, he modestly described himself as a 'labouring man'. Hanley was, in fact, a first-rate though entirely self-taught engineer with a keen eye for engine defects and problems, and long after his retirement his name was remembered and respected among NCC men.

Bowman Malcolm rarely travelled on the footplate, preferring to arrive unannounced at sheds around the system and talk to the men at random. They considered him a good engineer, appreciated his willingness to talk to them about their work, and were pleased at the excellent condition in which he maintained the locomotive fleet. Most engines were shopped after running 60,000 miles, which meant in practice every two years. He was strict in relation to alcohol, constantly warning his men about the evils of drinking. Enginemen who smoked on the footplate fared little better, as Malcolm reckoned that this gave the travelling public the impression of slovenliness!

Harold Houston, an engineer and enthusiast, followed his father into NCC service in 1920 and served as an apprentice during Malcolm's last two years in the Locomotive Department. He remembers that Malcolm arrived for work each morning at 8am and left at 5pm to catch a tram home – no company cars in those days, unlike today's pampered executives. During the day he would tour the works at precisely the same time in the morning and afternoon. If he saw anything displeasing, he turned about sharply and made for the foreman's office to complain. He would never allow overtime to be worked: his opinion was that if overtime was needed, either the man was too small for the job, or the job too big for the man!

Malcolm's railway interests did not end with the BNCR. He was a founder member of the revived Association of Railway Locomotive Engineers and was present at the first meeting in the St Pancras Hotel, London, on 1 January 1890, along with Martin Atock of the MGWR, HA Ivatt of the GS&WR and JC Park of the GNRI. For the rest of his career he was a regular attender at Association meetings, keeping abreast of railway developments in other companies.

At the 1897 meeting of the Association, Malcolm raised the matter of faulty stays and boiler explosions, one of which had recently occurred on the BNCR. The following year he became President of the ARLE. During the next two years the topic of firebox stays was raised again, McIntosh of the Caledonian Railway advocating bronze stays, though admitting the difficulty of riveting them. Malcolm hosted the 1906 summer meeting of the ARLE in the NCC's hotel at Portrush, at which the major topic for discussion was spark arresters. At the 1911 meeting it fell to Malcolm to propose REL Maunsell, then of the GS&WR, for membership.

Apart from a busy life on the railway, Malcolm was a governor of the Royal Belfast Academical Institution, one of Ireland's leading Public Schools. 'Inst' at this time was building a reputation as a rugby school, and would-be apprentices were always asked if they played the game. Knowing of Malcolm's connection with the school, they answered 'yes' – only to provoke an angry tirade as he denounced rugby as the last stronghold of barbarism in the Northern Hemisphere, and deprecated the number of working days lost due to the game!

As mentioned in Chapter Two, Malcolm began modestly with more of the established G class 2-4-0s and K class 0-6-0s, producing no new design until the F class 2-4-0s arrived in 1880. They was contemporaneous with the GNRI's H class 2-4-0s, also

a Beyer Product of similar dimensions. Tables 11 and 12 give the chronology and dimensions of this class.

The three F class 2-4-0s were considered the BNCR's crack engines, working the best Londonderry trains in the late 1880s and 1890s. Surviving mileage records indicate that Nos 45 and 46 covered over 1,600,000 miles each, while the slightly younger No 23 fell just short of this figure. No 45 became a Royal Train engine in 1885, transporting the Prince of Wales (later King Edward VII) from Larne Harbour to Belfast. In recognition of this, No 45 afterwards carried the Prince of Wales' feathers on her cab sides.

On one occasion in 1895 No 23, with a load of 95 tons, ran from Londonderry to Coleraine in a net time of 49 minutes, giving an average speed of 41mph – a creditable effort for such a small engine.

It was fortunate that the F class survived into the beginning of the modern railway enthusiast era in the 1930s, when Nos 23 and 45 were shedded at Coleraine and working the lighter Portrush branch trains. One brief snippet of performance has survived from this

Left: By 1880 the BNCR were in need of new passenger engines and Malcolm produced the first two F Class 2-4-0s. This is a Beyer Peacock 'Works grey' photograph of No 46. The tender is labelled 'BNCR' but the number plate is a painted copy.
LGRP

Below: A third F class engine, No 23, was built in 1885. She is seen at Belfast shed about 1895 in late BNCR livery. Note the sloping smoke box, parallel chimney and highly polished Beyer Peacock Works plate on the splasher.
LGRP

period: even with the light load, the speed reached between Portstewart and Coleraine was not often exceeded on the branch during the steam era.

Date	July 1937	
Loco	46	
Load	3 bogies [63/69tons]	
Portrush	**00.00**	min 23½ on the 1/76
Portstewart	**05.50**	*(bold indicates a stop)*
Coleraine	**04.43**	max 55

In 1938, No 46 returned to Belfast and spent her last year shunting the passenger yard. 23 stayed longer in Coleraine, returning to York Road late in 1938 to be used for shunting until 1940. The war, however, gained her a reprieve. Instead of being scrapped, she was shopped in July 1941 when the engine situation was desperate, and lasted until November 1942.

As shown in Table 11, No 46 (along with K class No 38) was fitted with a Phoenix superheater, and some comments on this contraption may be in order. Until the turn of the century engines were unsuperheated,

Left F class No 23, after rebuilding in 1899, is seen working hard on a Larne line train near Jordanstown in BNCR days. Note the clean and well kept permanent way. The path along the line, for the convenience of passengers, can be seen to the right of the engine behind the fence. The lamp code is correct for an ordinary passenger train.

Author's collection

Right: In 1910, No 46 was the subject of the tests with the Phoenix superheater mentioned in the text. It gave the engine a nose-heavy appearance and did nothing for the steaming. As the engine is in post-war MR NCC livery, the superheater had already been removed when this picture was taken.

Official NCC photograph

and steam generated in the boiler was described as 'saturated', because being in contact with boiling water it contained moisture. For this reason, older drivers often used the term 'wet steam'. The development of superheaters allowed wet steam to be passed through a series of tubes in direct contact with combustion gases, which raised its temperature and converted the moisture into even more steam. Superheated steam was of a sufficiently high temperature to remain above saturation level until after it had been used in the cylinders, resulting in more efficient boilers and a saving in coal of up to 25%.

The most reliable and effective type of superheater was developed by Wilhelm Schmidt, who placed the superheater in the boiler flue tubes – in the hottest part of the boiler. Steam from the boiler travelled through the regulator valve to the superheater header, where its flow was diverted into the superheater elements – long

multiple coils inside the 5½" diameter flue tubes, The superheated steam then returned to the main steam pipe and thence to the valve chests. In engineering, every improvement comes at a price and superheating, thus described, made cylinder lubrication more difficult. 'Wet steam' deposited a film of liquid on the sliding surfaces and this, aided by the steady trickle of oil from a hydrostatic lubricator, provided sufficient lubrication. A later refinement was the sight feed lubricator, where boiler steam forced oil through separate nozzles in drops which could be regulated by the driver. The coming of 'dry' steam eliminated condensation, and necessitated the development of better mechanical lubricators.

The foregoing describes superheaters as adopted by most railways, so what was the Phoenix superheater, and why did the BNCR use it? The simple answer was expense. The Schmidt superheater was expensive,

Left: William Robb has caught No 23 leaving Portrush on 18 June 1932 with the branch train of six-wheeled stock. William Robb, who sadly died recently, confessed to me once that BCDR moving trains could be coped with but GNRI trains moved much too fast for his camera!

William Robb

Below: A good tender-first view of an NCC 'bread cart' tender at Limavady Junction with a train from Dungiven. A number plate was carried and, if tenders were swapped the plates were also changed. Note the spare coupling hung over the left lamp iron.

Real Photographs Ltd

Opposite bottom left: The doyen of the F class, No 45 is pictured in final condition at Limavady about 1932, still with the small boiler and Ross pop valves. The lettering has been ground off the Beyer Works plate on the splasher and a tender weather board has been fitted for the Dungiven branch. Note the cattle wagon behind engine – at one time a prime earner for the NCC – and the brake rigging placed outside the wheels, to be more accessible.

LGRP

Opposite bottom right: In 1928 No 46 received a second hand 'five foot' boiler, becoming Class F1. She is seen at Larne in September 1932.

RG Jarvis, Midland Railway Trust, Ltd

covered by international patents which would have added over £50 to the cost of each locomotive. The Phoenix superheater, by contrast, was much cheaper, and in fact the initial fitting may have been free. Unfortunately, it was a total failure for several reasons. For a start, it was a nest of pipes in a smokebox extension and this upset the engine's blast. Next, the smokebox, unlike the flue tubes, is the coolest part of a boiler, and this guaranteed that there was little superheat to be had. Again, an engine smokebox forms the harshest possible environment; corrosive gases and accumulated ash made the superheater difficult to maintain. Finally, a superheater thus positioned added seriously to the weight of the front end.

Phoenix superheaters were tried for four years in No 38 and for about two years in No 46, before they were removed, though the engines retained the extended and disfiguring smokeboxes for many years after. The GNRI also experimented with the Phoenix superheater, though with just as little success. A

fine photograph of it appears on page 96 of another Colourpoint publication: *Locomotives of the GNRI* by Norman Johnston. One further question remains – why were Nos 38 and 46 the chosen engines for this device? The answer appears to be that, in 1910, superheating was seen as an economical alternative to compounding. It was reckoned that a comparatively cheap superheater applied to a simple engine would give all the economies which were claimed for compounds, as well as allowing boiler pressures – and hence maintenance costs – to be reduced. Sadly, this did not prove to be the case, and the moral of this tale for locomotive improvers was, and is, that it is not a good idea to fill a smokebox with gadgets!

Apart from the F class, the only new engines produced for the BNCR before 1883 were four J class 2-4-0 tanks for branch line working in the Belfast area, to Larne and Ballyclare. A brief chronology of the class and a summary of dimensions can be found in Tables 11 and 12. Two comments may be made about

Left: A period picture of great character, about 1910, showing J class 2-4-0T No 47 at Belfast in original condition. Note the ornate lamp behind the engine, the Beyer Peacock works plate on the bunker and the nicely burnished rods. The driver, in a soft hat, has a beard to rival WG Grace, while the fireman does not look happy clutching the long oiler.

Author's collection

Right: An action shot of No 25 leaving Belfast on the 9.45am for Londonderry on 16 September 1909. She will take it only as far as Greenisland, where, with a change of direction, a main line engine will take over at the other end. The train comprises straight-sided bogie carriages with a four-wheeled van lettered 'MR'. The view shows something of York Road before the platforms were lengthened and colour light signals replaced the handsome gantry of somersaults.

Ken Nunn Collection, LCGB

the dimensions of the J class. First, as built, the engines carried a ridiculously small amount of water – hardly enough for a 15 mile journey. Second, and by contrast, they carried a relatively large amount of coal.

The J class were gradually rebuilt with saddle tanks starting with No 48 in 1890, No 49 in 1891, then after a long gap No 25 in 1911 and No 47 in July 1914. The stated reason was to increase adhesion weight, but with a tractive effort of only 8,500 lbs and a 13 ton axle load on the coupled axles, the adhesion weight seems more than adequate. As built, the adhesion factor would have been about 7:1, comparing favourably with the GNRI's JT class 2-4-2T with a factor of 4.5:1. A more likely explanation emerges from a study of pictures of the engines before and after rebuilding. It is clear that examining and oiling the eccentrics and valve motion of a rebuilt engine was a much easier task than on an original side tank J class.

The J class worked passenger trains on the Ballyclare branch, almost exclusively, until about 1930, hardly a strenuous task hauling one bogie and a six wheeler! A rather more strenuous task sometimes came the way of the branch engine, and is illustrated opposite. This involved hauling a mainline train from Belfast to Carrickfergus Junction (later Greenisland) where a tender engine backed onto the rear of the train and took it forward to Londonderry.

Still on the lookout for engines, the BNCR bought a second-hand 0-4-2 (Sharp Stewart 2743) in 1886. This engine had a varied career, beginning in 1878 as No 5 of the Newry and Armagh Railway but, having placed their order, the N&AR paid only a quarter of the price, before finding that they could not really afford to buy the engine. The GNRI acquired the Newry and Armagh in 1879 but had no wish to settle its unpaid bills, so No 5 was sold to Mr Cooper, Chairman of the Belfast Central Railway in 1880. In 1883 it was overhauled by Grendons of Drogheda, and, in fact, carried a Grendons plate for the rest of its career. In the same year, Mr Cooper sold it to the Belfast Central Railway.

With what could be the same crew as seen on No 47 opposite, No 49 is seen as a 2-4-0ST at Belfast about 1911. She was rebuilt as a saddle tank in 1891. The tools used by the crew seem to be slung on every available handrail. The fireman looks gloomy but, seeing the size of the lumps of coal in the bunker, this is hardly surprising!

Author's collection

No 48 (rebuilt in 1890) at Belfast, well turned out in Midland livery with the crest on the cab side sheet. Note the hand grip beside the tank filler and the sandbox mounted above the footplating, in contrast to No 49, above. The brake block would certainly fail a test had such a thing existed in the steam age.

Author's collection

J class 2-4-0ST No 47 at Belfast shed about 1930, towards the end of its life. The headlamp code is for an ordinary passenger train. Note the screw link coupling hung over the right hand lamp iron. This was the last of the class to remain a side tank, not being rebuilt until 1914, though in 1932 she was the first to be withdrawn. Note the slight variation in the tank handrail brackets between this engine and No 25, below.

RS Carpenter,
Derek Young collection

Londonderry must have found a use for a small tank engine, as No 25 is seen here inside the shed in September 1932. The diamond at the chimney probably means she was working Limavady locals. The tablet catcher is mounted on the bunker, though No 25 can have had little use for it.

RG Jarvis, Midland Railway Trust, Ltd

In 1885 the GNRI absorbed the Belfast Central, and No 5 appeared in the GNRI's assets again. It was no more fortunate the second time around and, in the same manner as the fat boy who was never picked for the team, the Great Northern again put the engine on the market. This time Bowman Malcolm, scenting a bargain, bought the unwanted engine for £750.

The story of expensive engines and impecunious buyers was to recur in Irish locomotive history, finishing with the saga of the two Sligo, Leitrim and Northern Counties 0-6-4 tanks *Lough Melvin* and *Lough Erne*, which languished at Beyer, Peacocks' works from 1949 until 1951 while the SLNCR tried to find the money to pay for them. By coincidence both these engines featured in later NCC locomotive history, ending their days at York Road as UTA Nos 26 and 27. Indeed, No 27 *Lough Erne* remains on NCC metals to this day, albeit stored by the Railway Preservation Society of Ireland

at Whitehead.

Having added his bargain 0-4-2 to BNCR stock, Malcolm lost no time in putting it through the works and allocating it the number '50'. (It will be seen later that she was the first engine to carry this number, the second being the much more famous compound 4-4-0 No 50 *Jubilee*). Working all her days in the Belfast area, No 50 was renumbered '9' in September 1887 when the Sharp 2-2-2 of that number was scrapped and, briefly, '9a' in December 1904, before being finally scrapped in March 1905. Rarely can a £750 engine have given such good service! The mechanical details of this engine appear in Table 12.

By 1886 the BNCR owned fifty broad gauge engines. 2-2-2 Nos 9 and 21, 2-4-0 Nos 20, 23 and 34, 0-4-2 No 33 and 2-2-2WT No 24 – survivors from the B&BR, BBC&PJR and the L&CR – were finishing their days.

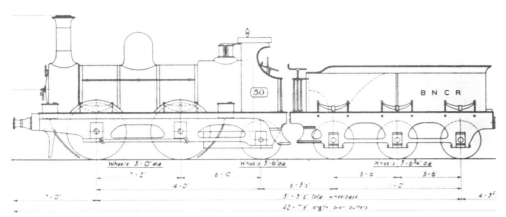

Sharp Stewart 0-4-2 No 50, purchased second hand from the Belfast Central Railway in 1886. Apart from having a six-wheeled tender, it was similar to the class built for the BCDR in 1880-90. In 1887 it was renumbered 9. Note the position of the 'BNCR' on the tender, similar to how it appears on the tender of 2-4-0 No 40 on page 31.

JH Houston

Nos 12–17 were the H class 2-4-0s for branch line and secondary duties. Nos 1–5 were the I class 2-4-0s, used mainly on secondary duties.

Nos 18, 19 and 35 were the L class goods engines and Nos 36 and 37 the L1 class. Nos 7, 28, 30–32, 38, 39, 43 and 44 were the K class 0-6-0s. These classes between them handled nearly all the goods traffic, and the larger wheeled K class did some passenger work as well.

Nos 6, 8, 10, 11, 22, 27, 29, 40 and 41 were the G class 2-4-0s, which worked the bulk of the passenger trains. F class 2-4-0 Nos 23, 45 and 46 worked the Portrush expresses.

Nos 25 and 47–49 were the J class 2-4-0 tanks for the

minor branches. This left such miscellaneous engines as 0-4-2 Nos 26 and 50, and No 42, the 0-4-0ST.

The fastest BNCR train in 1889 was a Belfast to Larne Harbour service which, including principal stops, did the run in 45 minutes at an average of 40mph. Sadly today no comparison can be made, as all trains now serve all stations, taking around an hour for an admittedly longer run from Belfast Central. The main line was less distinguished. The BNCR got a somewhat grudging mention in Foxwell and Farrar's "Express Trains, English and Foreign" for two Portrush trains which averaged 34 mph including stops or 37 mph without stops.

Table 11: Chronology of locomotives built 1880–89

Class	Type	No	Deliv	Builder	Cost	Rebuilt	Rebuilt	Rebuilt 2-4-0ST	Rebuilt LB	Scrapped
F	2-4-0	45	6/1880	BP 1921	£2120	1899	1912			9/1938
F	2-4-0	46	6/1880	BP 1922	£2120	1898	1907 1910		11/1928 (F1)	9/1938
F	2-4-0	23	5/1885	BP 2648	£2690	1899	1902			11/1942
J	2-4-0T	25	3/1883	BP 2233	£1954		6/1911*	1914		5/1934
J	2-4-0T	47	3/1883	BP 2234	£1954		7/1914*	7/1914		12/1932
J	2-4-0T	48	3/1883	BP 2235	£1954		1902*	10/1890		9/1933
J	2-4-0T	49	3/1883	BP 2236	£1954		1899*	9/1891		5/1934
–	0-4-2	50	6/1886*	SS 2743	£750 sh					3/1905

Notes: The F class were all reboilered in 1898–99, No 45 also receiving 17" cylinders.

Nos 23 and 46 got 17" cylinders in 1902/07.

No 46 received a Phoenix superheater and extended smokebox in 1910, the superheater soon removed and the extended smokebox later.

No 23 was initially withdrawn in Dec 1940 but reinstated in July 1941.

The J class carried a similar boiler to the I class 2-4-0s. * The first rebuilding column is for new boilers.

* 0-4-2 No 50 – This locomotive was originally built in 1878 and was bought second-hand. It was renumbered '9' in September 1887 and '9A' in December 1904.

Table 12: Dimensions of F and J classes and the BCR 0-4-2

Class/numbers	F class as built	F class final condition	F1 class No 46	J class as built	J class rebuilt	0-4-2 No 50
Type	2-4-0	2-4-0	2-4-0	2-4-0T	2-4-0ST	0-4-2
Cylinders	16"x22"	17"x24"	17"x24"	15"x20"	15"x20"	16"x22"
Coupled wheels	6' 0"	6' 0"	6' 0"	5' 2⅝"	5' 2⅝"	5' 0"
Leading wheels	4'1¾"	4'1¾"	4'1¾"	3' 8"	3' 8"	–
Trailing wheels	–	–	–	–	–	3'6"
Wheel base	7' 1½" + 7' 3"	7' 1½" + 7' 3"	7' 1½" + 7' 3"	7' 2" + 7' 6"	7' 2" + 7' 6"	7' 2" + 6' 10"
Boiler length	10' 0"	10' 3½"	9' 11"	9' 10"	9' 10"	10' 2"
diameter	4' 1"	4' 0"	4' 8¼"	3' 9"	3' 9"	3' 10"
tubes	152 x 2"	161 x 1⅞"	193 x 2"	132 x 1⅞"	134 x 1⅞"	172 x 1⅞"
Heating surface	850 sq ft	860 sq ft	937 sq ft	770 sq ft		955 sq ft
Firebox (inner)	3'8⅞" x3'10⅞"		3'8⅞" x3'10⅜"	3' 6" x 4' 0"		4'0" x 3'10⅜"
Grate area	14½ sq ft	14½ sq ft	14½ sq ft	14 sq ft	13¼ sq ft	15½ sq ft
Boiler pressure	150 psi	160 psi	170 psi	140 psi	140 psi	140 psi
Tractive effort	9973 lbs	13,101 lbs	13,920 lbs	8551 lbs	8551 lbs	11,170 lbs
Weight	32¾ tons	35 tons	38 tons, 1 cwt	35 tons, 8 cwt	37 tons, 8 cwt	38 tons
Water capacity	1456 gallons			500 gallons	750 gallons	1500 gallons
Coal capacity				1 ton, 17 cwt	1 ton, 17 cwt	

Chapter 4
The Compound Years

No 33 *Galgorm Castle*, the first compound, in the official works photograph in 1890 – Beyer, Peacock No 3058. The nameplate on the splasher is combined with the Beyer, Peacock makers plate. The tender has two very large tool boxes and the lever reverse is prominent. No 33 was the first engine in the British Isles to have inside Walschaert's valve gear and the first compound to run in Ireland. Embracing compounding elevated Bowman Malcolm from being an unknown engineer of a minor railway to the forefront of his profession. *LGRP*

By 1889 the BNCR was once again in need of new engines, and Malcolm approached Beyer Peacock with details of his requirements. On previous occasions, more F class 2-4-0s would have sufficed but, coincidentally with the BNCR's needs, the idea of compounding was becoming popular around Europe. An important figure in this part of the story is Herman Lange who, after engineering experience in Berlin and Karlsruhe, came to England to work for Beyer Peacock. He became chief draughtsman with Beyer in 1865 and Chief Engineer in 1876 following Charles Beyer's death, the same year that Malcolm took charge at York Road. Lange was closely interested in compounds and, as Michael Rutherford has recently pointed out*, he took out a patent in 1888 to improve the starting valve on the Worsdell-Von Borries system of compounding. Two years later, Lange was instrumental in persuading Malcolm to try a two-cylinder compound 2-4-0 of about the size of an F class. An order was placed in November 1889 and the first engine of what became the C class was delivered in April 1890. This was No 33, quickly followed by Nos 50–52

A word might not be out of place on the ordering and building of engines at this time. Traditionally, it has been assumed that a new engine was designed by the engineer of the purchasing company, and for the larger

companies this was certainly so. But smaller companies like the BNCR would not have had a large enough staff to produce a complete set of drawings for a new engine in their own drawing office. It is most likely that Malcolm would have indicated to Beyer Peacock the size and type of engine required, and then taken their advice on its design. During the erection of the engine, a BNCR representative would have been in attendance to see that satisfactory standards were maintained and, following the engine's delivery, a representative of the builder (and a set of drawings) would have come with the engine and stayed during its running-in period.

Table 13 gives the basic chronology of this class. An interesting point is that the building costs of each locomotive decreased slightly between 1890 and 1895. The 1890s were clearly a cheap period to stock up with new engines. Indeed, the last new BNCR engine of comparable size, F class No 23 of 1886, cost £2690, whereas C class No 57 cost £2450 nearly a decade later.

Table 14 shows the basic dimensions of the C class. Inside Walschaerts valve gear was fitted – the first time this was used on an inside cylinder engine – and the slide valves were inclined over the cylinders. This valve gear gave excellent steam distribution at all points of the cut off, and was easily and economically maintained.

* 'Railway Reflections No 123', in *Backtrack*, September 2006

The original No 51 of 1890 seen as an oil burner between 1896 and 1901. She is in full BNCR 'invisible green' livery with the BNCR crest on the cab. The brass splasher beading has 'Beyer Peacock' and the building date. The very short seven-foot wheelbase is clear in this picture and parts of the Walschaerts motion can be made out.

Real Photographs Ltd

Photos of No 21 with her original number are rare. (She became No 51 in 1928.) She is seen here at Belfast on 22 May 1924, still lettered 'MR' with many railway men getting in on the picture. A 4-4-0 compound can just be seen behind her.

Real Photographs Ltd

The subsequent mechanical history of the class, and details of their changes of number, make an interesting but highly complex study. No 33, the first of the C class, and the first BNCR engine since 1851 to carry a name, was named *Galgorm Castle* after the home of the BNCR general manager. She was scrapped in 1926 (although some parts may have gone to U1 class 4-4-0 No 3) and the name lay in abeyance until 1931 when it went to No 57, before vanishing in 1938.

No 50 became 58 in 1895, so that the new 7'0" compound *Jubilee* could take the appropriate number 50. Near the end of her career, when rebuilt in 1927 as a B3 class 4-4-0, No 58 was renumbered 28, receiving the name *County Tyrone* in 1932.

No 52 received a second-hand 'five foot' boiler in 1928, becoming Class C1. In 1931 she was withdrawn but contributed some parts to new U1 class 4-4-0 No 4.

No 51 was fitted with Holden's patent oil burning apparatus from November 1896 until 1901. In 1926 she received a bogie and second-hand 'five foot' boiler, becoming a B1 class 4-4-0. After a more thorough rebuild in 1928 as a B3 simple, she received the number 21, and in 1932 was named *County Down*.

Another compound, No 21, was built in 1892. She was rebuilt with a 'five foot' boiler in 1928 and took the number 51, in an identity swop with the engine just described! The logic behind this was to have all C class in the fifty series and the B3 4-4-0s in the twenties.

The last two C class compounds Nos 56 and 57, built in 1895, were the longest-lived, No 56 surviving in original condition until November 1942. No 57 was the first engine in the world to be fitted with the Ross 'Pop' safety valve, named after its Coleraine inventor. Ross had the first valve made in York Road shops and fitted

This view at Londonderry was taken about 1898 and shows C class No 50, in original form, after her 1895 renumbering as '58'. Note the toolbox perched precariously on top of the coal.   *Derek Young collection*

In the 1930s Cookstown shed relied almost entirely on compounds for its turns – the last two-cylinder compound depot in the British Isles, possibly in the world. This picture allows comparison between C1 class No 51 with the large boiler and C class No 56 with the original type. Both engines are in LMS red livery.   *LGRP*

it experimentally to No 57. When she was scrapped in 1938 the valve was preserved in the drawing office, and miraculously rescued from the debris by Harold Houston when York Road works was severely damaged in the German blitz of 1941. No 57 carried the name *Galgorm Castle* from 1931 until withdrawal in 1938. Nos 56 and 57 were, therefore, the last 2-4-0s built for the BNCR, No 56 also being the last surviving NCC 2-4-0 at the time of her scrapping in 1942.

It might be useful to reflect at this point on the fate of other Irish 2-4-0 engines of the period. BCDR No 6 of 1894 was the last 2-4-0 built for the County Down.

She was easily the last 2-4-0 to run in Northern Ireland, working until January 1950, although not sold for scrap by the Ulster Transport Authority until 1956. The last Great Northern 2-4-0s were H class Nos 84–87, built in 1880–81. They were disposed of as early as 1932, making the GNRI the second Irish railway to eliminate this wheel arrangement.

Further south, in 1925 the Great Southern Railways inherited 2-4-0s from four constituents. The solitary Dublin South Eastern example (Class G7) and the six Great Southern and Western G4 class had all been scrapped in 1928. However, the remainder – four

Waterford and Limerick G3 class and 19 excellent G2s from the Midland Great Western Railway – survived into CIE days and, in the case of the latter, some to the 1960s. (Twenty MGWR 2-4-0s would have been inherited, had not No 20 been destroyed during the Civil War in the Irish Free State.)

To the MGWR engines fell three distinctions. First, they were the last 2-4-0s to be built in Ireland, the last of them being delivered in 1898. Secondly, they were probably the last 2-4-0s in the world to be regularly employed on express trains, working the Sligo to Dublin night mail train until 1956. Finally, they became the last 2-4-0s to survive in Ireland, CIE No 653 lasting until 1963.

In weight and power terms, the Midland engines were about the same as the BNCR C class. BCDR No 6 was also very similar before its rebuilding in 1943, and at first glance could have been mistaken for a BNCR compound. Indeed, the County Down had three C class of their own, built in 1892, having borrowed a BNCR C class for comparative trials before they placed their order with Beyer Peacock. However, in 1894, when the BCDR needed a fourth engine, they went for a simple version of the compound design to give more power.

Few people now remember the sight of a 2-4-0 at work on any Irish railway, and fewer still can remember back over 60 years to the BNCR C's. How sad that neither the UTA in the 1950s nor CIE in the 1960s could see their way to preserving even one example of such a distinguished and distinctive nineteenth century locomotive type!

Most of the NCC records were destroyed during the 1941 blitz, but some information has survived on the C class 2-4-0s, along with the memories of some of the men who worked on them. No 33 was remembered for most of the period from 1910 to 1930 as a Cookstown engine, in the hands of drivers Blair and Murray. She just failed to run a million miles in service, the records indicating that she was withdrawn after 952,211 miles. No 51 (originally No 21), on the other hand, did cross the million mark, accumulating 1,062,076 miles during the long and complex career described above. On 12 August 1926, No 52, at that time a Coleraine engine, found herself at the head of a heavy Apprentice Boys special of bogie and six-wheeled stock after U1 4-4-0 No 1 failed. Although the train was theoretically beyond her capacity, little time was booked against her.

No 21 was another engine to spend time in Coleraine. With her regular driver Doyle she put up a fair mileage between Coleraine and Londonderry, though the Derry portions of mainline trains rarely exceeded four lightweight bogies. Both engine and

Above: This close up shows the front end of No 57 in 1937, with the large low pressure cylinder dwarfing the high pressure cylinder on the left. The differing sizes were intended to achieve equal work from cylinders working at different pressures.
HC Casserley

Left: A delightful picture of C1 class No 51 blowing off at Portrush in the late 1930s, with the overall roof visible in the background. The driver is enjoying the afternoon sunshine as he awaits departure with probably a Derry Central train, possibly the 1.40pm. The NCC used standard LMS 'Period 1' corridor stock and a side-corridor third is at the adjacent platform.
Derek Young collection

driver feature in a story which may be apocryphal, but which does indicate something of the pride that older men took in 'their' engines. Booking on duty one day, Doyle found that his regular engine was not awaiting him. He stormed into the foreman's office to complain, only to be told: "There's a heavy train to be worked today, and it's sharply timed. So I thought you'd better leave your engine here, and take one of the Company's instead." This from a canny foreman who knew that Doyle would never abuse his beloved No 21.

As noted earlier, No 56 lasted longest, ending her days unrebuilt in 1942 with a mileage of 1,203,589. After a spell in Larne, she became a stalwart of Cookstown shed, and her last days were spent shunting at Coleraine with occasional trips down the Portrush branch. Her most arduous task was the steeply graded branch to Coleraine Harbour. Going down to the Harbour, it might almost be said that the wagons took the engine, but coming back was very different. With a heavily loaded train of ammunition, a clear run into the station was needed, as a signal check before the station level crossing would result in the train dragging the engine back down the branch. If this happened too often, the fireman would sometimes gently heat one end of the large brass staff (one of only seven exceptions to the NCC's near-universal use of the tablet for single line sections) before innocently handing it down to the signalman as they passed the cabin.

No 57 had a varied career, including a spell in Larne when she merited a small footnote in 20th century Irish history. During the Home Rule crisis of 1912–14 Winston Churchill, then Home Secretary, visited Belfast on 7 February 1912 to address a meeting of the Ulster Liberal Association. He received a less than cordial reception from the Unionist community, and eventually had to be taken back to Larne in a special train, worked by No 57, to board the Stranraer ferry.

Of more importance to the railway enthusiast is 57's later career as a Cookstown engine. She was noted regularly in the mid 1930s on through workings

Right: C1 class No 51 at Cookstown about 1938, preparing to leave with either a Belfast train or, perhaps, the 9.15am to Portrush. Cookstown had two stations, side by side, and was one of two places where the GNRI and NCC met (Antrim being the other). An enthusiast for the 2-4-0 in the late twenties would have been well served at Cookstown as the GNRI engine was also a 2-4-0. No 51 was the former No 21, the two engines having exchanged numbers in 1928 when the original No 51 became a B3 4-4-0.

Real Photographs Ltd

Left: Another view of No 51 as a C1, this time at Coleraine with a train for the 'Derry Central'. A smart looking No 51 has recently been through the shops and received a full re-paint into LMS livery.
The Derry Central line left the NCC main line at Macfin, proceeded over the Bann on a fine girder bridge and ran through Aghadowey (my grandfather's last station), Garvagh and Kilrea to Magherafelt.

Author's collection

over the Great Northern branch from Cookstown to Dungannon and during 1936 was a regular on the 9.15am Cookstown–Portrush. Normally composed of three lightweight bogies, this train could load up to six on a fine summer Sunday. On arrival at Portrush, the fire was cleaned and No 57 took the 1.10pm from Portrush to Derry, finishing off with the 5.25pm Londonderry–Belfast dining car express as far as Coleraine. Although only a three bogie train as far as Coleraine, the 5.25pm was booked to run the 15½ miles to Limavady Junction in 20 minutes, including a conditional stop at Eglinton. For the 18 miles from the Junction to Coleraine, this train was booked for 35 minutes including four stops; sadly no logs exist. Neither has any photograph survived of the unique sight at Coleraine of a two-cylinder compound of 1895 vintage standing beside the virtually new mogul which would have brought the main portion of the train through from Portrush to be combined for Belfast.

The E class 0-6-0s

After two years experience with compound 2-4-0s, Malcolm next applied the same principle to 0-6-0 goods engines. In April and May 1892, E class Nos 53 and 54 were delivered. Their builders numbers were 3457 and 3458, they cost the company £2736 each, and they survived until May 1934 and September 1944 respectively. The dimensions of this class appear in Table 14. As originally built, these engines were to have received 18"/26"x24" cylinders, but the order was changed to 17"/25"x24" before delivery. No 53 received the larger cylinders in February 1911 and a 'five foot' boiler and larger cab in October 1921. No 54 received the 'five foot' boiler in October 1907 and the new cylinders in January 1914. In 'five foot' boiler form, No 53 had a large firebox and was classified 'E1'.

Rated as an immediate success, these engines went straight on to the Derry goods, which they worked day and day about. No 53's first regular crew were driver Rafferty and fireman Kealy and she was normally based at Belfast, though she spent time at Ballymena after finishing on the Derry goods. Here she did some passenger work in the 1920s, and on 12 July 1930 she is recorded as taking a special from Maghera to Portstewart. No 53's duties also included a spell on the Cookstown 'shipper'. Her mileage on withdrawal was 1,126,578.

Right: Official Beyer Peacock photograph of E class No 53. The light compounds having proved successful, it was natural that Malcolm should try the principle on a goods engine. Nos 53 and 54 were built in 1892. Note the two large toolboxes on the rear of the tender.

Beyer, Peacock

Left: No 53 at Londonderry, not long after delivery, with the brass dome and brass ring of the safety valves burnished. The large tool boxes on the back of the tender have gone.

Real Photographs Ltd

Opposite bottom: No 53 sitting on the shed road at Ballymena on 9 August 1930. Since 0-6-0s usually had no footsteps at the front of the running plate, the fireman is standing on the coupling rod to examine the inside Walschaerts gear.

HC Casserley

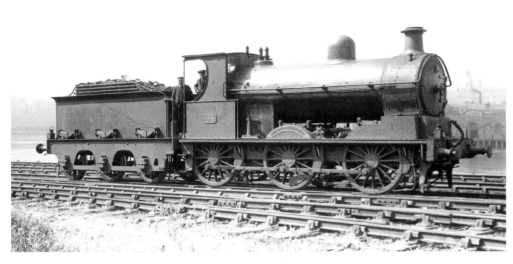

A nice study of No 54 sitting beside the river at Londonderry on 15 May 1937. On the other bank is the GNRI, which handled more goods traffic with much bigger engines than No 54. Part of the inside motion can be clearly seen. Ross 'pop' valves replaced Ramsbottom on both engines, probably when reboilered with 'five foot' boilers in 1907 (54) and 1921 (53).

RG Jarvis
Midland Railway Trust, Ltd

No 54 was the Derry-based engine, and her original crew was driver James Dennison and fireman James Logue. She outlasted her sister by some ten years, running about 1,400,000 miles. The latter part of her life was spent shunting at York Road, though in 1938 she covered nearly 7000 miles on ballast trains, and double that figure in 1939. She did nothing in 1940, but wartime conditions earned her a reprieve and she re-entered service in January 1941, to run another 66,000 miles before final withdrawal in 1944. By this time she was the last compound goods engine in the British Isles.

Nos 53 and 54 had only a steam brake on the tender, and 54 was usually reckoned the better steaming engine of the pair. Out of Ballymena in the down direction they could take 36 wagons but, beyond this load, the Ballymena pilot, with the banking staff, would assist to Cullybackey. (Most NCC brake vans were only seven tons weight and, since the banker was not to be coupled to the train, a goods train with a 'pilot load' was required to have two brake vans on the rear.)

Jubilee and *Parkmount*

By 1895 the first five C class 2-4-0s and the two goods engines were performing satisfactorily, and Malcolm decided on an enlarged express engine. For the new D class, the 2-4-0 wheel arrangement was retained, but the driving wheel diameter of 7'0" was the largest ever used in Ireland. In Victorian times the concept of speed was associated with wheel size, (an idea which was finally debunked when the British Railways 9F 2-10-0s showed what they could do with only 5'0" wheels!). Apart from Churchward, who was thirty years ahead of his time, there was little understanding of the relationship between front-end design and efficient free running. There is in fact no performance evidence to suggest that the 7'0" engines were any faster than their 6'0" sisters, but the BNCR believed them to be the fastest engines in Ireland, and so a legend was born. Perhaps, at a time when the average height of a man was about 5'4", an engineman gazing up at a seven foot driving wheel would have had an awesome impression of power and therefore speed.

The two D class 2-4-0s were Nos 50 *Jubilee*, and 55 *Parkmount*, and their builders numbers were 2632 and 2633. The naming of the former had nothing to do with Queen Victoria, whose Jubilee was in 1887. It was so named because 1895 was the fiftieth anniversary of the incorporation of the Belfast and Ballymena Railway in 1845 and C class 2-4-0 No 50 was renumbered to allow the number to be transferred to *Jubilee* to match the name.

The two engines went into traffic in June 1895, costing the company £2690 each. They ran as 2-4-0s for only a short time, being altered to 4-4-0s in 1897. No 55 survived

After the compound goods, two heavier express engines were ordered, Nos 50 *Jubilee* and 55 *Parkmount*. The Works photo of No 50 in photographic grey suggests an engine that will not be steady at speed. The 7'0" driving wheels are clearly obvious when compared with the Works photo of No 33 on page 47 and the 8'2" wheel base emphasizes the size of the wheels.

Beyer, Peacock

Unsteadiness in No 50 must have shown up early in the engine's career, as this rather poor, but rare, photograph shows a redesign of the front springs about 1896, presumably in an early attempt to improve the poor riding. It must have been unsuccessful as, in 1897, both were rebuilt as 4-4-0s.

BNCR

until 1944, and No 50 until 1946. No 50 ran as a compound until 1926 and was then converted to a simple engine with a small superheated boiler. No 55 was never rebuilt, finishing her days as perhaps the last two cylinder compound to run on a main line in the British Isles. The dimensions of these engines can be found in Table 15.

It is claimed that both engines were altered to 4-4-0s due to the speeds they attained, but a look at the engines as originally built would suggest a more plausible explanation – they were too heavy at the front end. As 2-4-0s they weighed 43 tons, which made them the heaviest 2-4-0s in Ireland by a margin of nearly six tons. The C class and the BCDR 2-4-0s weighed 37¼tons and the Midland 2-4-0s 37½tons. After a redesign of the front springs (see photo above) failed to solve the problem, Beyer Peacock was approached on this matter, and the leading bogie which was eventually fitted to Nos 50 and 55 was in fact standard with that used by Beyer at the time.

Turning 2-4-0s into 4-4-0s both solved the weight problem at the front end and, as a bonus, gave the engines a most impressive appearance. It also increased the weight to 46 tons, but this was reasonable by the

standards of contemporary 4-4-0s. They weighed a ton more than the GNRI PP class, which was probably a stronger engine, and were much heavier than the GSWR 60 class which were a type noted for speed. On the other hand, the BNCR D class had strong frames and good axle boxes, whereas the PP and 60 classes were more lightly built to satisfy their Civil Engineers. A lesson was learned and, apart from the two C class 2-4-0s, Nos 56 and 57, which arrived just after 50 and 55, every new passenger engine for the following 35 years had a bogie. The mighty seven foot wheel was, sensibly, abandoned.

As might be expected of Malcolm's prize engines, *Jubilee* and *Parkmount* went straight on to the 6.30am Derry express, which conveyed the mails, and the 3.30pm afternoon express. In those days, before the direct loop line was built, there was a change of direction at Carrickfergus Junction (later Greenisland) and any available engine hauled the train to the Junction. Here the express engine came on to the rear and took it up the old main line to Monkstown and thence to the North West.

There are some scanty surviving records of No 55's work. In 1907 she was working that morning's Breakfast

Top: Even that authoritative magazine *The Engineer* seems to have been impressed with *Parkmount* as they produced this beautiful woodcut of her which I used for making my Gauge O model of her.

The Engineer

Below: A nice study of D class No 50 *Jubilee* as a 4-4-0 showing the extremely handsome lines of this engine.

Author's collection

Car express from Greenisland to Londonderry, allowed two hours 35 minutes, with five stops. A rather sketchy log suggests that, with 190 tons, she ran from Carrickfergus Junction to Ballymena in 38 minutes and 12 seconds, and from Limavady Junction to Derry in 20 minutes. In the up direction, with 120 tons, she ran from Derry to Coleraine in 40 minutes, but no reliable details survive.

Fortunately, however, the late Billy Hanley had personal memories of No 55 on the 3.30pm express:

"The 3.30 was a hard train for the long-legged, but not very strong, compounds to time. They had the reputation of being slow up the bank but fast downhill. Comfortable to ride on, they had a slow rolling motion which never got any worse. They were worked on nearly full regulator. The 3.30 carried slips for Ballyclare and Cookstown, and 55 would be thrashed from Greenisland to Ballyclare Junction to gain sufficient speed to drop the first slip. This would put the water about out of the glass, and it was after Doagh before the boiler was right again. Speed built up to over 70 at Muckamore, and after easing for the curve the engine would again be worked hard to accelerate back to 70 at Niblock Crossing (milepost 23½, at the

foot of the bank beyond Antrim). The second slip was at Cookstown Junction (milepost 25) – if speed was high enough, and they had enough way on it. But if the train wasn't moving fast enough a stop would have to be made at Cookstown Junction, the Rules requiring paper work to explain why!"

Slip working broke one of the strictest rules of railway safety by having two trains in one block section. It was carried out only by senior guards, and never in fog or snow. Signalmen had to ensure that the slip coach had come off the main train and reached its destination before giving 'line clear'. The slip coach

D class No 55 passes Whiteabbey with a special train made up of six-wheeled coaches. A slotted-post signal is beside the fourth carriage and the path between stations for convenience of passengers is on the left of the picture.

LGRP

No 50 *Jubilee* leaving Ballymena on an up express. Two signals are pulled off, the one to the right being for a narrow gauge Larne train. The traditional, but frowned upon, race between narrow and broad gauge is taking place and on this occasion *Jubilee* is the clear winner with the narrow gauge not in sight. Bill Hanley saw the 'race' as a small boy and described it as "a terrier chasing a greyhound". The race was unpopular with Management because the three foot gauge turned sharply left not far from Ballymena.   *Author's collection*

carried two tail lamps, one red and one white, carried side by side, and also side lamps. The rear vehicle of the main train, of course, carried a tail lamp.

When she was superseded on the Derry expresses by the newer A class compounds before World War I, No 55 moved to Ballymena, where she spent many years. She was reboilered in 1908 and 1924, and ran her biggest annual mileage in 1907–08. By the early 1930s she had moved to Cookstown, acquiring a weatherboard on the tender to work the Draperstown branch. 1938 saw her in Belfast again, sometimes covering for railcar failures on the Larne line, and occasionally on the 8am from York Road. This was a most interesting train which went to Killagan, of all places, before running round and returning to Ballymena as a feeder into the 8.15am from Londonderry. It seemed that 1938 was her swan-song, for in 1939 she did little work, and none at all in 1940. As with other engines, the war procured a reprieve for *Parkmount*, and after a shopping she soldiered on until 1944, finishing her life having run 1,301,142 miles.

There is an amusing story about that last visit to the works during the war. Around this time, an officious NCC accountant complained to Malcolm Patrick about engines which existed "only on paper": Patrick took some delight in bringing the accountant into the works to view 'the paper engine'.

No 50 ran 1,025,914 miles as compound, but in 1926 she was rebuilt as a simple with 19"x24" cylinders and

a small superheated boiler (ex-No 70). In this form she ran a further 335,528 miles until 1938, when it seemed that her days were numbered. The threat of war saved her life, however, and she emerged from the works in 1939 with a reconditioned boiler from V class 0-6-0 No 13. She ran a further 200,000 miles, finishing her days in 1946 with 1,561,442 miles to her credit.

Nos 50 and 55 were quite unlike any other NCC engines in having Crewe-look cabs and open coupling rod splashers, both features retained in No 50 when she was rebuilt. When No 50 emerged from the works in 1939 in black livery, she looked like a Crewe engine – albeit with a Derby chimney. Perhaps the most handsome engine to run on the NCC, No 50 was classed D1 after being rebuilt as a simple.

No 50 made a small piece of NCC history on 21 March 1924, working the 2.30pm Coleraine to Londonderry, which was the first service train to cross the new Bann Bridge at Coleraine. After her years on the expresses, No 50 became something of a gypsy on the system. She was a Larne engine in the 1920s, working boat trains and Carrickfergus locals, and by 1931 she could be seen each evening working the 6pm Derry express out to Greenisland, returning with the 3pm from Derry. During the 1933 strike, when white collar staff kept some services going, Harold Houston drove her on the Boat Express. This train was allowed 35 minutes to Larne Harbour, and if the boat was up to five minutes late arriving, the engine was expected to make the time up! Harold remembered this as one of his most satisfying railway experiences, taking No 50 along the shore from Whitehead on the 30 minute timing, which was no problem with four or five coaches.

In the early 1930s she went to Coleraine, becoming a popular performer on the Coleraine–Derry section, as well as the Limavady branch. She later spent time in Belfast and Ballymena, handling various workings including the peculiar local service which ran from Cullybackey to Antrim and then on to Aldergrove on the GNRI Antrim branch.

By the end of the 1890s Malcolm was well satisfied with his eleven compounds, but new engines were still needed. Thus a further five Beyer Peacock compounds arrived – Nos 59–62 in 1897 and No 24 in 1898, identical to the C class but with a leading bogie and a larger tender. The chronology of what became the B class is contained in Table 13, and basic dimensions in Table 15. The Bs were the first compound class to vanish

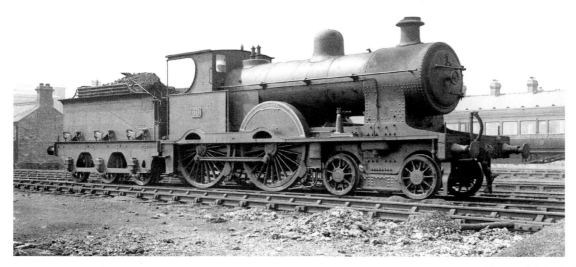

Above: Jubilee's final rebuilding did nothing to spoil her looks. Indeed, Harold Houston thought her the most handsome NCC engine. In this photo she looks almost like a product of Crewe. The Fowler-Anderson bypass valve is visible below the footplating – this was supposed to help the engine coast more freely. She has a poor looking load of slack in the tender.

RG Jarvis,
Midland Railway Trust, Ltd

Above: No 50 with the 8.35am to Belfast, at Larne Harbour on 11 July 1931. As well as the main subject of the picture, there are fine examples of mixed gauge track in the sidings in the foreground. The narrow gauge passenger trains used the other side of the island platform.

Real Photographs Ltd

Right: *Parkmount* passes Mossley on the 1pm up Cookstown train on 15 September 1933. Single line working was in force as the loop line was under construction. The train make up is interesting – a van and two open wagons at the front then two bogies and two more vans.

William Robb

In the author's opinion, the B class compounds were the most handsome engines to run in Ireland. This is the official works photograph of No 62. Note the 'BNCR' in beautiful scroll work on the splasher.

LGRP

B class 4-4-0 No 60, with what is almost certainly the Royal train from Londonderry to Newtownards in 1897. The train comprises a six-wheel van at each end, two bogie saloons at the front and two six-wheel saloons at the rear.

Author's collection

A lovely study of No 62 at Portrush in MR NCC days, with the MR coat of arms on the cab side. The NCC-style balance weights are prominent on the driving axle – the rear drivers seem to need little balancing.

Author's collection

completely and, because of their early withdrawal, few records remain of their work which, by the 1920s, was mostly on the Larne line. Fortunately, one log does survive of a boat train in February 1928 on what, for that time, was a very sharp timing of 35 minutes. Sadly there are now no non-stop scheduled runs from Larne to Belfast to test NIR's new CAF railcars, but in comparative terms No 60 did not disgrace herself in relation to either modern traction or the famous 30 minute steam timing of the 1930s.

Loco: 4-4-0 No 60	Load: 100/110 tons	
	Time	*Speed*
LARNE HARBOUR	**00.00**	
Larne Town	04.00	
Slow start round the long curve to Larne Town		
Glynn	05.40	53
Magheramorne	08.10	
Ballycarry	12.10	49
Whitehead	14.45	39
Kilroot	19.10	48
Carrickfergus	21.40	36
Signal check, followed by sound climb of Mount bank		
Trooperslane	24.30	40
Greenisland	26.35	49
Jordanstown	28.00	60
Whiteabbey	29.10	63/66
BELFAST	**34.45**	
Fast run up from Greenisland		

Surviving memories of these engines are scanty but interesting, and some of these are mechanical. NCC compounds could not normally be worked as simples, as a disc intercepting valve changed the engines to full compounding after the first revolution of the wheel. This could make starting troublesome, and so No 59 was tried with an experimental cylinder fitted to the interceptor valve. This gave high pressure steam to both the low and high pressure cylinders and enabled the engine to work as a simple until the train was accelerating. Unfortunately steam was rapidly exhausted as the boiler tried to feed a 23¼"x24" cylinder, and the idea was quickly abandoned.

No 24 was altered in 1925 with two simple 18"x24" cylinders and a 'five foot' boiler, which increased her weight to 42½ tons and her tractive effort to 14,300 lbs. Surprisingly, the large saturated boiler was quite unable to supply enough steam, and No 24's limp performances were a frequent cause of line blockages. She would frequently come to a stand at Greencastle – some two miles from Belfast – with neither steam nor water and, as this was a seaside location, the engine acquired the nickname of 'the diver'.

No 62 had the misfortune to run into a landslide at Briggs' Loop at milepost 13½ on the Larne line whilst working the 6.30pm up Boat Train on 19 February 1910. The engine and all seven vehicles became derailed, though there were no serious injuries. As she remained upright, No 62 was trailed back up to the line along a temporary track laid for the purpose, and the line reopened to traffic within two days. One wonders how long it would take to deal with a similar incident today!

The fifth member of the B class, No 24, at Portrush in 1899, bearing the BNCR scrollwork on the splasher. BNCR/NCC engines carried their numbers on the tender as well as the cab side, so each engine required three plates. Note the toolbox, which featured in the Works photo, still perched high on the tender.

Author's collection

The handsome lines of the B class are well captured in this view of No 62 at Belfast in early Midland Railway days. The white diamond suggests she is being prepared for a Larne line working, although the head lamp code indicates an express train. The tender has what looks like slack in it.

Author's collection

In 1921 Nos 60 and 61 were rebuilt as Class B1 with 'five foot' boilers, giving them a bulkier appearance. No 60 is seen here at Belfast in 1928. She was completely modernised in 1932 as a superheated simple and became class B3.

Kelland Collection, Bournemouth Railway Club

Nos 24, 59 and 62 retained the small boilers. Here, No 59 is seen at Portrush with an experimental cylinder fitted to the interceptor valve alongside the smokebox (see page 59).

Author's collection

No 24 was converted to a simple in 1925 with a 'five foot' saturated boiler and high-sided tender, neither of which improved the looks of the engine. In this form she was classified B2, but the rebuild was unsuccessful and she was further rebuilt in 1928 with a G6S superheated boiler to Class B3.

Author's collection

A nice study of No 60 waiting to depart from York Road Platform 1, shortly after rebuilding to Class B1 in 1921. Wright's Coal Tar Soap is advertised prominently on the building in the background.

Derek Young collection

Table 13: Chronology of compound locomotives built 1890–98

Class	Type	No	Deliv	Builder	Cost	Rebuilt	Renumbered	Renewed	Scrapped
C	2-4-0	33	3/1890	BP 3058	£2505	1913		11/1926 (U1 No 3)	–
C	2-4-0	50	3/1890	BP 3059	£2505		3/1895 (58)	6/1927 (B3 No 28)	–
C	2-4-0	51	5/1890	BP 3200	£2505	8/1926 (B1)		12/1928 (B3 No 21)	–
C	2-4-0	52	6/1890	BP 3201	£2505	8/1928 (C1)		1/1931 (U1 No 4)	–
C	2-4-0	21	12/1892	BP 3521	£2550	10/1928 (C1)	10/1928 (51)		9/1938
C	2-4-0	56	7/1895	BP 3680	£2450				11/1942
C	2-4-0	57	7/1895	BP 3681	£2450	11/1931 (C1)			9/1938
E	0-6-0	53	4/1892	BP 3457	£2736	2/1911 (cyls) 10/1921 (E1)			5/1934
E	0-6-0	54	4/1892	BP 3458	£2736	10/1907 (cyls & 4' 8¼" blr)			9/1944
D	2-4-0	50	5/1895	BP 3632	£2690	1897 (4-4-0) 4/1926 (D1)			10/1946
D	2-4-0	55	6/1895	BP 3633	£2690	1897 (4-4-0)			9/1944
B	4-4-0	59	6/1897	BP 3885				5/1924 (U1 No 1)	–
B	4-4-0	60	7/1897	BP 3886		4/1921 (B1)		6/1932 (B3)	–
B	4-4-0	61	10/1897	BP 3887		7/1921 (B1)		1/1932 (B3)	–
B	4-4-0	62	11/1897	BP 3888				7/1924 (U1 No 2)	–
B	4-4-0	24	12/1898	BP 4059		2/1925 (B2)		9/1928 (B3)	–

Notes: C class No 33 was named *Galgorm Castle* until renewal in 1926.

C class No 57 was the first locomotive in the world to receive 'Ross pop' safety valves.

C class No 57 was named *Galgorm Castle* in December 1931, using the plates from No 33.

E class No 54 was the last two-cylinder compound 0-6-0 in the British Isles. No 54 was initially withdrawn in September 1940, but reinstated in January 1941.

D class No 50 was named *Jubilee*. In 1926 it rebuilt using the boiler from No 70. It was given a similar boiler from V class 0-6-0 No 13 in 10/1939.

D class No 55 was named *Parkmount*. It was the last two-cylinder compound 4-4-0 in the British Isles.

Table 14: Dimensions of C, C1, E and E1 classes

Class/numbers	C class	C1 class	E class	E class as rebuilt	E1 class
Type	2-4-0	2-4-0	0-6-0	0-6-0	0-6-0
Cylinders	16/23¼"x24"	16/23¼"x24"	17½/25"x24"	18/26"x24"	18/26"x24"
Coupled wheels	6' 0"	6' 0"	5' 2⅝"	5' 2⅝"	5' 2⅝"
Leading wheels	4'1¾"	4'1¾"	–	–	–
Wheel base	7' 6" + 7' 0"	7' 6" + 7' 0"	7' 5" + 7' 10"	7' 5" + 7' 10"	7' 5" + 7' 10"
Boiler length	10' 0"	9' 11"	11' 0"	10' 0"	9' 10"
diameter	4' 1"	4' 8¼"	4' 1"	4' 8¼"	4' 8¼"
tubes	152 x 1⅞"	193 x 2"	152 x 1⅞"	193 x 2"	193 x 2"
Heating surface	861 sq ft	936 sq ft	936 sq ft		956 sq ft
Firebox (inner)	3'8⅞" x3'10⅜"	3'8⅞" x3'10⅜"	3'8⅞" x3'10⅝"		
Grate area	14½ sq ft	14½ sq ft	14 sq ft	14½ sq ft	18¼ sq ft
Boiler pressure	170 psi	170 psi	170 psi	170 psi	170 psi
Tractive effort	12,330 lbs	12,330 lbs	16,004 lbs	17,942 lbs	17,942 lbs
Weight	35 tons, 10½ cwt	39 tons, 17 cwt	39 tons, 8½ cwt	43 tons, 11 cwt	?
Water capacity	1440 gallons	1440 gallons	1440 gallons	1440 gallons	
Coal capacity	5 tons	5 tons	6 tons	6 tons	

Notes: Between 1911 and 1921 Class E No 53 had larger cylinders, but retained the 4' 1" boiler.

Table 15: Dimensions of D, D1, B, B1 and B2 classes

Class/numbers	D class	D1 class	B class	B1 class	B2 class
Type	2-4-0/4-4-0	4-4-0	4-4-0	4-4-0	4-4-0
Cylinders	18/26"x24"	19"x24"	16/23¼"x24"	16/23¼"x24"	18"x24"
Coupled wheels	7' 0"	7' 0"	6' 0"	6' 0"	6' 0"
Leading/bogie wheels	4'1¾"/3' 0"	3' 0"	3' 0"	3' 0"	3' 0"
Wheel base	7' 8" + 8' 2"	6' 6" + 6' 7" + 8' 2"	6' 6" + 6' 6" + 7' 0"	6' 6" + 6' 6" + 7' 0"	6' 6" + 6' 6" + 7' 0"
Boiler length	10' 4⅞"	10' 4¼"	10' 0"	9' 11"	9' 10"
diameter	4' 4"	4' 4⅛"	4' 1"	4' 8¼"	4' 8¼"
tubes	199 x 1⅞"	127 x 1¾"/18 x 5⅛"	152 x 1⅞"	193 x 2"	193 x 2"
Heating surface	1144 sq ft	1261.5 sq ft	850 sq ft	936 sq ft	
Firebox	4'8⅞" x3'10⅜"	4'8⅝" x3'10⅜"	3'8⅞" x3'10⅝"	3'8⅞" x3'10⅝"	
Grate area	18¼ sq ft	18¼ sq ft	14 sq ft	14½ sq ft	14½ sq ft
Boiler pressure	170 psi	170 psi	160 psi	170 psi	170 psi
Tractive effort	13,685 lbs	14,904 lbs	11,560 lbs	12,282 lbs	15,606 lbs
Weight	43¼/45½ tons	46 tons, 12 cwt	39 tons, 17 cwt	43 tons, 8 cwt	42 tons, 8 cwt
Tender weight	28¾ tons	28 tons, 17 cwt	28 tons, 17 cwt	28 tons, 17 cwt	28 tons, 17 cwt
Water capacity	2090 gallons	2090 gallons	2090 gallons	2090 gallons	2090 gallons
Coal capacity	6 tons	6 tons	6 tons	6 tons	6 tons

Chapter 5
The A Class: An Appraisal of the Von Borries Engines

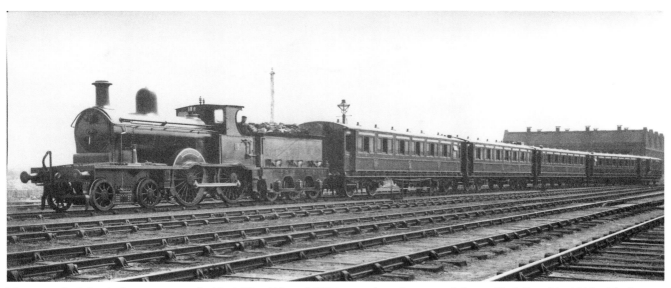

The second of the A class compounds, No 3 *King Edward VII*, is seen here in a posed shot at York Road in 1903 with the Holden tour train. There is no screw coupling on the front of the engine, just a single link, which became an NCC feature. The engine is still in BNCR green and the paintwork is so polished that the tablet catcher is reflected in it. The Holden train was the idea of a Larne hotelier, who ran summer tours of Ulster. The all-first coaches were built by Malcolm at York Road and were a great credit to the workmanship of the York Road shops. Since the engine was also built at York Road, the whole ensemble was a fine example of what the BNCR could do. *Official BNCR photo*

Numbering thirteen in total, the A class was the most numerous class of compounds on the NCC, and indeed the largest class on the railway until the delivery of the fourteenth U2 class 4-4-0 in 1934. The next time this number was equalled was in 1942 when the fourteenth member of the fifteen-strong W class of 2-6-0 arrived. Both the U2 4-4-0 and WT 2-6-4T classes eventually numbered eighteen.

Sixteen broad gauge compounds to four designs were in service by 1901 when Malcolm designed his largest and last class of engines. A chronology of the A class appears in Table 16 and the dimensions can be found in Table 17. The first two, Nos 34 *Queen Alexandra* and 3 *King Edward VII* were the only ones to be named. They were the first engines to be built at York Road for nearly thirty years and with them, as Malcolm himself admitted, the ultimate in size had been reached. In a two-cylinder compound, the double expansion of steam requires a large low pressure cylinder. The more powerful the engine, the larger becomes the cylinder, until a point is reached where the engine can no longer be contained within the British loading gauge.

The appearance of the A class also marked the end of two eras. They were the last engines to be built for

the BNCR before the Midland Railway takeover, and their construction ended the long relationship between the NCC and Beyer Peacock. B class No 24 was the last Beyer engine delivered to the BNCR, bringing to an end a period of thirty-five years in which 33 engines had been purchased.

Table 16 contains the first reference to Derby, which was to become another major player on the NCC locomotive scene. After their takeover of the BNCR in 1903, the Midland Railway of England naturally began to take an interest in the locomotive policy of their acquisition. Although Malcolm realised that the limit had been reached for two-cylinder compounding, he managed to convince his new masters that the A class was ideal for the NCC and, so, construction continued at York Road and six of the class were even built at Derby. It is worth noting that, at this stage, the Midland were building their own three-cylinder compounds – surely the only British locomotive works ever to have simultaneously built two and three-cylinder compounds for two different gauges! Also worthy of note is the significant variation in prices over the eight year building span, with Nos 5 and 17 in 1906 and 1907 the cheapest engines, and No 4 in 1903 the dearest.

Before commenting further on the engines, something should be said about a unique experiment in Automatic Train Control in Ireland which was tried on No 17, the last Belfast-built broad gauge compound. ATC was pioneered in England by the Great Western Railway on the line from Oxford to Fairford via Witney, and was an electro-mechanical system activated by the aspect shown by the distant signal. The NCC system was entirely mechanical, designed in 1922 by a Belfast engineer, Andrew Kerr, and was intended to show the aspects of both distant and home signals. If the distant signal was 'off' a siren sounded in the cab, and if it was 'on' a horn sounded, followed by a partial brake application. If the home signal was clear there was no sound, but if it was at danger a full brake application was triggered to the accompaniment of a continuous horn. This full brake application could only be cancelled when the engine stopped and the driver left the footplate and pressed a knob on the apparatus, which was adjacent to the vacuum brake pipe on the front buffer beam. The trials in 1922 seem to have been successful, but the system progressed no further due to maintenance problems with the large number of moving parts in the apparatus, which was quickly removed from No 17.

The A class was distributed around the system, and by the summer of 1915, engines 3, 4 and 34 were at Ballymena. Whilst shedded here, No 34 was regularly cleaned by young Billy Hanley, supervised by head cleaner Ned Gillespie. On one occasion they carelessly left No 3's regulator slightly open, with No 34 behind her in a similar state. As pressure built up in the receiver chest, No 3 moved forward and then stopped again, but not before No 34 had also moved forward and run into her. Little damage was done, but the terrific noise of the impact brought the Foreman from his office, and earned Hanley and Gillespie three days' suspension

each! No 4 was a regular engine for the 3.30pm ex-Belfast, with the Cookstown slip portion. For most of her life No 5 was a Belfast engine and worked on the main line until 1924, crewed by drivers Campbell and Nixon for a record 27 years. One of her turns was the 6.30am to Londonderry, returning on the 11.10am.

One day in 1917, No 5 was working the Saturday 12 noon 'Golfers Express' when she shed part of a driving tyre at Dromona Siding (milepost 38½). A G class 2-4-0 was sent out from Ballymena to pull the train back to Cullybackey, while No 5 ran cautiously on, light engine, to Glarryford. The missing part of the tyre was not found until No 5 arrived back at the works, where it was discovered on top of the brick arch! Almost unbelievably, it had come up through the ashpan and into the firebox without causing damage! Some years later, by rare coincidence, the same engine shed a tyre on the same wheel near Ballymena Goods while working the 5.40pm ex-Londonderry, which connected with the Heysham Steamer. Ballymena supplied another compound, and the train was far less delayed than would be the case if a similar accident befell a train in the twenty-first century!

Even this did not satisfy No 5's desire to shed tyres. This time she was working the 9.55am down mail train from Belfast (which conveyed the Coleraine TPO) with Billy Hanley and Jimmy Keenan. As Hanley applied the brake on the way down Ballyboyland bank he instinctively noted that it did not act as quickly as usual, and almost immediately a violent knocking began. When they stopped, Keenan walked forward and found that two feet of tyre, on exactly the same wheel, was missing. This time one piece had stripped part of the engine's brake gear and broken off a washout plug, leaving only the threaded part, before embedding itself in the embankment. Another fragment of metal

(Continued on page 66)

A class No 4 was built at Belfast in 1903 and is seen about 1907-08. The works plate is obscured by the jack. Note the steam sanding equipment and short cab.

Author's collection

No 34 *Queen Alexandra*, the first of the A class compounds, photographed in the traditional place at Portrush. The first of the Belfast-built engines, steam sanding has been fitted. The oval Belfast builder's plate can be seen in front of the jack. Very few BNCR engines were named and those that were had the name on the splasher rather than above it.

Author's collection

In 1904, A class No 9 appeared from York Road, the first to be built under Midland Railway auspices. She is seen here in the later MR NCC style, with the crest on the splasher and 'NCC' in scroll on the tender. Note the snifting valve mounted on the smokebox. No 9 became 69 in 1925 in one of the NCC's re-numbering schemes, when as many as possible were given numbers in the 58–69 range.

Author's collection

In 1905 a batch of four A class compounds (Nos 63–66) were built at Derby for the NCC. No 65 of this batch is seen at Belfast about 1911. Ross 'pop' safety valves were fitted to this class from new. The engine still has two whistles – one being for 'crowing'. This class had 'cast in' balance weights rather than the NCC 'bolt on' type.

Author's collection

In the Edwardian years it was the practice to built one A class 4-4-0 each year at York Road works. No 20 was built in 1905, followed by No 5 in 1906. The latter is seen pausing at Larne with a goods for Belfast about 1920. The imposing bracket signal still has conventional arms, rather than the later somersault type. The line to Larne Harbour curves to right behind the engine. No 5 became 59 in the 1925 renumbering

Real Photographs Ltd

A class No 65 again, this time on a ballast train near Cookstown Junction on 4 May 1920. Note the well-kept permanent way. It seems strange to see an A class compound on a ballast train at a time when there were plenty of 0-6-0s.
Ken Nunn Collection, LCGB

(Continued from page 64)

passed under the engine and tender.

Leaving the fireman and guard to put down detonators front and rear, Hanley walked forward to Glenlough crossing, telephoned Ballymoney for assistance, and then borrowed a bike to cycle on to Ballymoney with the tablet, so that the section could be cleared for an assisting engine from Coleraine. Hanley and Keenan then put the disgraced No 5 into the Quarry Siding at Ballyboyland, and eventually arrived in Coleraine with their train, reckoning they had done a reasonable bit of work in the circumstances. Not a bit of it! The first person to meet them was the shed foreman, very grumpy that they hadn't tried to bring No 5 light engine to Coleraine: "Sure you still had seven wheels left on the engine . . . !" When No 5 eventually got to Belfast she was given a completely new wheelset, and this seemed to cure her. Hanley, however, maintained that she was still the roughest of the class.

Other members of the class had a less dramatic history. After 1912, No 17 spent most of her life in Londonderry, while No 63 finished her days as a Cookstown engine, at a period when that shed had an entirely compound allocation including C class Nos 51, 56 and 57.

In April 1927, No 65 was tried as a simple with two 18"x24" cylinders and piston valves. She was successful enough with the existing boiler, but no further conversions were made as it was decided to modernise the whole class with new superheated G6S boilers, the first candidate being No 68 in December 1927.

No 68, in June 1908, was the last broad gauge engine to be built to a BNCR design (though constructed at Derby), and indeed the last new broad gauge engine to appear before 1914. Four more narrow gauge

compounds were yet to be constructed, as explained in chapter 11, two in 1908–09 and two as late as 1919–20, so the compound era was not quite over. Other than two railmotors, nothing but compounds had been built since 1890 – thirteen As, five Bs, seven Cs, two Ds and two E class totalling 29 engines out of a total broad gauge stock of 70 in 1908. In general, goods traffic was mostly worked by simples, and passenger trains by compounds, and this situation lasted until 1914. Why was it, then, that the Company placed so much emphasis on compounds, and what appraisal can be made of the compound years? The rest of the chapter deals with these questions.

The first question which arises is why Malcolm allowed himself to be persuaded to invest so heavily in the compound system. The answer can only be that he thought that its use would make economies in the biggest single item in the locomotive department budget – coal. In a speech to the Institute of Mechanical Engineers he claimed that the introduction of compounding had saved the BNCR 12% in coal consumption in the decade 1890–1900. He made similar claims in an interview printed in *The Railway Magazine*, quoting an impressive array of statistics which are worth examination.

He began with the First Belfast Passenger Link – the top link of its day, which covered the Derry and Portrush expresses – on which a saving of 14.2% was claimed for the C class compounds over the F class simples. For the Carrickfergus passenger link an even bigger saving of 17.8% was achieved by comparison between the compounds and the much older G and H classes of 1872 and 1856 respectively. The Carrick trains had smart timings, and the older engines had to be thrashed hard to keep them, whereas the newer compounds would have had something in hand. A similarly large saving of 11.2% was claimed in the Coleraine passenger link, the comparison here being with the capable F class. In the goods link, a very high figure of 16.5% was claimed, possibly reflecting 53 and 54's long tenure of the Derry goods.

Figures like these were music to the ears of the company accountants, but against fuel economies had to be balanced increased boiler maintenance costs arising from higher boiler pressures. To obtain economies from a compound, and get something like equality of work between the cylinders, higher boiler pressures were needed, and from the 150psi of the F class, the BNCR went to 160psi and then 175psi with the compounds. It would therefore seem that the 'economies' which Malcolm claimed were illusory, in that they were not comparing like with like.

Malcolm further maintained that the reduction of back pressure, along with only half the number of beats per revolution, softened the draught and gave more economical combustion. Because of the softness of the blast, less fire-cleaning was required, less ash accumulated in the smokebox and fewer sparks were emitted.

The late OS Nock, however, had a footplate run on a two-cylinder compound on the 4.20pm Cookstown train with engine No 57. She was worked away from stations on full regulator and in full forward gear, which would roughly equate with 25–30% cut off in a simple engine. He recollected that the two exhaust blasts from the low pressure cylinder became quite tremendous, so perhaps Malcolm's argument was not totally accurate!

By contrast with FW Webb's unpredictable compounds on the LNWR, which often required pilot engines, the BNCR's engines did all that was required of them for almost 25 years and gave the traffic department no cause for complaint. They seemed to be on top of their work and, unlike the aristocratic Webb, Malcolm was only too willing to discuss his engines and give regular interviews and lectures on their work. There could, however, be truth in the late

Above: Those A class that had still not been rebuilt by the LMS, usually remained in invisible green after 1923. No 69 (ex-9) is seen at Londonderry in 1931. The peculiar hoist behind the engine is possibly for coal (see also page 33). *LGRP*

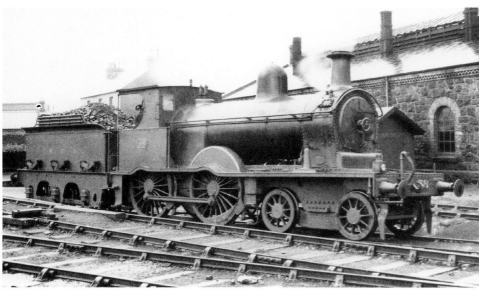

Left: No 58 (ex-17) at Coleraine in 1932. The A class were known as the 'heavy compounds' as they were about six tons heavier than the B class though slightly less heavy than *Parkmount*. Judging by the damp shed roof, Coleraine is experiencing a typical wet day. *RG Jarvis*

The Midland Railway Trust, Ltd

In 1932 the nameplates *Queen Alexandra*, removed from No 34 when it was rebuilt in 1928, were given to No 63. This picture probably dates from 1933–35, as she was renewed as U2 class 4-4-0 No 87 in 1936, retaining the nameplates.

Real Photographs Ltd

Harold Houston's opinion that he persisted with them for rather too long.

A contributory factor to the compounds' success in traffic was that BNCR trains were light. Coaches in the 1890s were typically only 48' 0" long and weighed around 22–25 tons. Since third class coaches could accommodate between 70 and 80 people, a train weighing less than 120 tons, well within the capacity of a compound, could carry a lot of passengers.

Among enginemen, the compounds were a popular engine. They steamed well, particularly before the 'five foot' boilers were fitted, and the Walschaerts valve gear gave a nice distribution of steam. They were easy to drive, the change over to compound being automatic after the first revolution of the wheels. The lever reverse adjusted the valve travel in both high and low pressure cylinders simultaneously, although moving the lever took a considerable physical effort. For this reason, a steel plate was welded to the front of the reversing rack: putting one foot against it enabled the driver to get more pull at the lever!

Four notches took the engine to the full forward position which was used for starting. On rising gradients the engine was worked two notches back, which corresponded to 61% cut off in the high pressure and 68% in the low pressure cylinder. The regulator would be about seven-eights open for normal working though, when coasting, would be pulled back to the first valve – the 'wee valve' to NCC men – with the cut-off set four back to give 48% in the high pressure and 54% in the low pressure cylinder. The valve gear was arranged to give a longer cut-off in the low pressure cylinder as an aid to starting.

There were two significant problems, though. First, the compounds, particularly the B and C classes, were not confident starters, and often stopped in such a way that only the small high pressure cylinder got steam as the engine tried to restart. Experienced drivers trying to avoid this problem could be seen hanging out of the cab coming into a station to watch the position of the coupling rod as they brought the engine to a stand. One particularly hard restart for down Larne trains was from Whitehead – straight up the 1 in 105 to Slaughterford Bridge before falling to Ballycarry. One day No 50's crew failed several times to achieve a start, setting back a bit further each time until eventually the whole train was out of the platform in the up direction. It did not help tempers on the footplate when the signalman leaned out of his cabin and asked if they wanted to give him back the tablet for Ballycarry, and get the Kilroot one back again!

There was one further problem. While the steam pressures in the high and low pressure cylinders balanced each other, there was a much greater weight of moving parts on the low-pressure side. The result was that the compounds were unbalanced engines, though this did not seem to inhibit their speed.

If, according to the well-known adage, imitation is said to be the sincerest form of flattery, then it would appear that no other railways flattered the NCC by rushing to build compounds on the same scale. The Great Northern, with a policy of continuous development of rugged types, declined Beyer, Peacock's offer of two-cylinder compounds in 1897. Thus, at the same time as Beyers were building compounds for the BNCR, they were also producing the Great Northern

Q class 4-4-0's (a natural progression from the P and PP classes), which in turn was developed into the S class, which worked main line expresses almost to the end of the steam era.

The Great Southern and Western at this time was under the firm grip of Ivatt, soon to move to the Great Northern of England. He experimented by converting No 93, one of the speedy '60 class' 4-4-0s, and No 165, one of the ubiquitous '101 class' 0-6-0s, from simples to compounds. Cylinder sizes were similar to the BNCR 'A class' compounds, and Stephenson motion was retained. Ivatt redesigned No 93's starting valve so that the engine could be worked in simple expansion when required, while leaving No 165 with the standard type of flap intercepting valve. Although he claimed in 1900 that No 93 was neither better nor worse than her simple sisters, it seems clear that she was not a successful engine, mainly because the influx of steam into the low pressure cylinder choked the exhaust passages. A further weakness related to boiler pressures. The pressures of the converted engines was left at 150psi – Inchicore was no believer in high pressure, and almost to the end of steam most of their older passenger and goods engines did not blow off above 160psi. Significantly, James Holden of the Great Eastern Railway reported at this time that the GER compounds working at 160psi had a 14% economy over simple engines, but that this figure reduced to only 2% with pressure reduced to 150psi.

The neighbouring BCDR had seven compounds. Three were 2-4-0s, virtually identical to the C class and, before placing the order, the County Down borrowed a BNCR 2-4-0 compound and worked it to Newcastle. After observing its lively behaviour, the Civil Engineer lobbied for a bogie engine, but was over-ruled. The County Down compound 2-4-0s worked mostly on the main line, and all had gone by 1921.

The other four engines, 2-4-2Ts built for the Bangor line, were much less successful. Joseph Tatlow and the BCDR board were fascinated by the novelties of compounding and anxious to have these engines for the summer of 1891. However, due to full order books, Beyer, Peacock could not meet this deadline and the BCDR had to wait until the end of the year. RG Millar, the Chief Mechanical Engineer, did not want them at all, as he was particularly aware of the starting problems inherent in the Von Borries system. While the antics described earlier at Whitehead station could just about be tolerated on the BNCR, they were quite unacceptable on the Bangor branch with its high density timetable, sharp gradients and frequent station stops. Millar also correctly reckoned that the compound tanks would be too rigid on the curves and too heavy for the line. They were indeed heavier than the 2-4-2 simple tanks which followed them, and as heavy as another of their successors, the handsome 'wee standard' 4-4-2 tanks. Some of these faults were solved with the substitution of leading bogies in the late 1890s, converting them to 4-4-2Ts, but the starting problem remained. Overall, these engines became a nuisance, and were offered for sale in 1921. Failing to find a buyer, the County Down eventually sold them for scrap.

As an aside, it must be concluded that the BCDR Board were slow learners, as this was not the last instance of their buying totally unsuitable engines. Shortly before the end of World War 1, and allegedly after being rather well wined and dined, the County Down directors were shown one of the legendary London, Brighton and South Coast Baltic tanks. They immediately decided that the County Down must have similar engines, and although Millar would have had the seniority to stand up to his directors, he was on the verge of retirement. His successor Crosthwait could not resist their folly, and was saddled with a severely pruned version of the Brighton giants. Although large and impressive, they were expensive to buy, voracious in their appetite for coal, and poor steamers. This digression from BNCR compounds actually leads back, in a convoluted way, to the NCC. In the dying days of steam on the Bangor line, the UTA sent over five of the superb new WT class 2-6-4 tanks – Nos 4, 7, 10, 50 and 53 – for varying periods in the early 1950s. They were everything the Baltics were not – strong, fast and economical, and the BCDR men took to them at once. Sadly, by 1953 the Bangor line was totally dieselised and the 'Jeeps' returned to York Road.

Summarising the compounds is a matter of weighing pros and cons. They were economical, a quality applauded by the accountants, but possibly exaggerated by their designers. Their efficiency and reliability kept the traffic department happy. Their free steaming propensities and good maintenance made them popular with footplatemen. Although, as we have seen, they could be difficult engines to start, especially on suburban stopping trains, drivers eventually mastered their peculiarities. Perhaps the fairest conclusion is that of EL Ahrons (author of the six volume *Locomotive and Train Working in the Latter part of the Nineteenth Century*, Cambridge 1954), "They were good engines for the times, but the times were of the Victorian era". And indeed, the NCC's typically lightweight trains prolonged the Victorian era into Edwardian times.

Table 16: Chronology of A class compound 4-4-0s

No	Name	Deliv	Builder	Cost	Renumbered	Renewed A1	Renewed U2	Mileage
34	Queen Alexandra	4/1901	York Rd	£3160		4/1928	–	870,932
3	King Edward VII	7/1902	York Rd	£2843	3/1926 (33)	*12/1932	–	952,211
4		8/1903	York Rd	£3080	9/1924 (62)	7/1928	–	780,212
9		12/1904	York Rd	£2761	9/1925 (69)	*6/1933	–	889,821
20		1905	York Rd	£2760	–	–	12/1929 (84)	n/a
63		5/1905	Derby	£2882	–	–	1/1936 (87)	947,930
64		5/1905	Derby	£2882	–	8/1929	–	823,444
65		5/1905	Derby	£2882	–	6/1929	–	870,966
66		5/1905	Derby	£2882	–	*5/1930	–	701,000
5		8/1906	York Rd	£2500	6/1925 (59)	–	9/1934 (86)	927,878
17		1/1907	York Rd	£2500	6/1927 (58)	*2/1934	–	832,099
67		6/1908	Derby	£2772	–	–	1/1934 (85)	855,721
68		6/1908	Derby	£2772	–	12/1927	–	617,447

Notes: The nameplates on No 34 were transferred to No 63 in November 1932, and retained after its renewal.

No 65 was experimentally rebuilt as a simple, with 18"x24" cylinders and piston valves, in April 1927.

* 200 psi boilers, others 160 psi

Table 17: Dimensions of A class compound 4-4-0s

Type	4-4-0
Cylinders	18/26"x24"
Coupled wheels	6' 0"
Bogie wheels	3' 0"
Wheel base	6' 6" + 6' 7" + 8' 2"
Boiler length	10' 4⅜"
diameter	4' 4¼"
tubes	199 x 1⅞"
Heating surface	1152 sq ft
Firebox	4'8⅞" x3'10⅝"
Grate area	18¼ sq ft
Boiler pressure	175 psi
Tractive effort	16,065 lbs
Weight	45 tons, 1 cwt
Tender weight	28 tons, 17 cwt
Water capacity	2090 gallons
Coal capacity	6 tons

Notes: Number of tubes later reduced to 182

The first two engines had 18¼" dia HP cyls

Chapter 6
Life under the Midland: The Derby Influence, 1903–1923

On 1 July 1903, after forty years of independent existence, the BNCR became the Midland Railway (Northern Counties Committee). Since the new NCC was run by a Belfast-based management committee, the takeover at first made very little difference. Engine livery remained the same olive green, though the letters 'MR NCC' began to appear on tender sides. Malcolm stayed in the chair at York Road and built another eleven A class compounds, five at York Road and six at Derby. Derby also built spare boilers for the heavy compounds.

The Derby engines came in two lots, the first of which, consisting of four locomotives, arrived in 1905 as order Nos 0 2833 (the engines) and 0 2839 (the tenders). One report suggested that the Derby boilers had 199 brass tubes, and the boilers pressed to 190psi (the pressure used in some Midland compound boilers). This is not, however, confirmed elsewhere. In 1907 order 0 3385 was placed for the remaining two compounds – the last BNCR inspired broad gauge locomotives and the last to be built before 1914.

Realising that the day of the compound was over, Malcolm next prepared a new design for a simple 4-4-0 with a superheated boiler as standard. This was the U class, a major advance on the compounds, and in 1914 two engines and tenders were built to orders 0 4369 and 0 4370 respectively. The dimensions of the U class appear in Table 19, and the chronology in Table 18.

Nos 69 and 70 came in 1914, with Nos 72 and 73 following in 1922, the time lag (and the significant increase in cost) being explained by the intervening World War I. The 1922 engines originally carried the numbers 14 and 15, but Malcolm's successor WK Wallace did some renumbering, and in March 1923 they became Nos 72 and 73.

U class No 70 leaving Belfast on 4 May 1920 under a gantry of semaphore signals with the 8.10am to Carrickfergus. Note the baffles under the gantry. Apart from the post-war Midland livery, No 70 is still largely in original condition with the short cab, as seen in the frontispiece on page 4. Although built at Derby in 1914, the U class did not use a standard Midland boiler, instead getting a superheated version of that on the 'heavy compounds'.
Ken Nunn Collection, LCGB

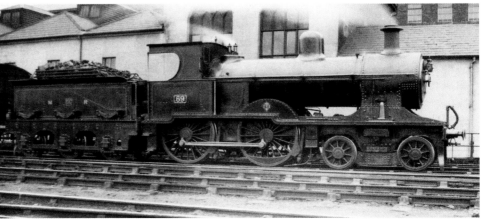

In contrast, No 69, which became No 71 in early 1923 had already received an extended cab roof. It is seen at Belfast, probably about 1922. Some passenger engines appeared at this time with the Midland crest on the splasher, rather than the cab side.
Author's collection

Top left: A rare picture of brand new U class No 73 during the brief period when she initially ran as No 15. The Fowler/Anderson by pass valve under the running plate is evidence of Derby influence.

Author's collection

Top right: U class No 72 on a Portrush train at Limavady Junction around 1930. The LMS had moved into the period of using concrete for platforms and walls.

Real Photographs Ltd

Centre upper: No 73 at Londonderry in 1932. The steeply rising ground behind the engine, together with the river on the other side, made the station a very constricted site.

RG Jarvis,
Midland Railway Trust, Ltd

Centre lower: V class 0-6-0 No 13 at Belfast on 5 August 1930. Called the 'heavy goods' by NCC men, they were really only the LMS equivalent of 3F. Although built in 1923, their tenders were still the old 'bread cart' type.

HC Casserley, courtesy RM Casserley

Right: The V class, as built, had the same round-top superheated boilers as the U class. Two, Nos 13 and 15, had boiler changes in 1938 when they got the boilers previously carried by Nos 72 and 73. No 15 is seen here outside York Road works awaiting attention, probably in 1938. The screw coupling hung over the lamp iron suggests she may have been shunting the passenger yard. No 14 did not get a boiler change and was the first to receive a second hand G6S Belpaire boiler in 1951, as seen opposite. Nos 13 and 15 did not receive theirs until 1953.

LGRP

Malcolm's last piece of design work for the NCC was the V class superheated 0-6-0, a goods version of the U class passenger locomotives, and identical to them in most respects, other than the wheel arrangement. A chronology of the class appears in Table 18. Derby Works produced none of the standard locomotives envisaged by the newly-formed LMS in 1923, and the V class, being already under construction, were among the first engines produced by the LMS after the amalgamation. The driving wheels were 5'2½" and the wheelbase 7'5"+7'10", the latter flouting an established Derby convention of 8'0"+8'6" for six coupled engines. The cylinders were inclined at 1:9 and driven by 8" piston valves with Walschaerts gear, and the engines weighed 46¾ tons, slightly lighter than the U class.

V class numbering was rather confusing. When they first appeared in January 1923 they were numbered 71, 72 and 73, but when the renumbering scheme was started York Road was unsure what to number them and they temporarily became X, Y and Z! In April 1923 they reverted to the dignity of numbers as 13, 14 and 15. Nicknamed the 'heavy goods' engines, they could show a fine turn of speed and were extensively used on excursion traffic to Portrush and local trains on the Larne line, frequently being timed at up to 60mph.

In the first ten years of their lives, the class was kept busy. No 13, at Cookstown Junction and No 15, at Whitehead, did as much local passenger as goods work, while No 14 at Belfast was regularly on the 7.45pm Ballymena goods with drivers McAllister and Gracey. This train was banked to Ballyclare Junction, where it shunted to be overtaken by the following Ballymena passenger train.

By 1935, the engine fire boxes were wearing out, and the effects of the 1933 strike and the depressed years of the 1930s were taking their toll. No 13 was stored out of traffic between 1935 and 1937, and No 15 between 1937 and 1938. No 14 got a new firebox in 1935, and Nos 13 and 15 were thoroughly overhauled at York Road in 1938, receiving reconditioned boilers out of U class Nos 72 and 73, which had been rebuilt as U2 class in 1937. During the Second World War

Above left and right: The UTA had a poor reputation in many regards but its lined black livery could look most attractive and these two pictures show No 14 at Belfast in September 1951, just after rebuilding with a G6S boiler to class V1.

Both author

Left: The last regular work the V1 class had was over the Cookstown and Derry Central lines. No 13 is sitting at Cookstown Junction with the goods from Kilrea in 1959. To the right can be seen the main line, and beyond it a siding with cattle wagons, once an important traffic for the NCC. Behind the engine is the shed, with the water column coming out of the wall behind the bread container.

A Donaldson

Top: The up Kilrea goods is seen here in 1956 behind No 14, which is taking water at Magherafelt, at one time the junction for Coleraine, Cookstown and Draperstown. The railway still had some cattle traffic at that time.

A Donaldson

Below: Watched by the ticket collector, No 14 arrives with a Larne line local at Berkeley Dean Wise's beautiful Swiss chalet style station at Trooperslane, now sadly replaced by a metal shelter.

RM Arnold

No 15 was shopped by the GNRI at Dundalk in 1943 and later ran on the GNRB from January to May 1956.

The V class were the last engines with some BNCR influence in their design. Bowman Malcolm retired in September 1922, aged 68, after 46 years in charge at York Road – a reign all the more remarkable in that from 1906 he had also been responsible for the Civil Engineering department. In that role he had a major role in the design of the new Bann Bridge at Coleraine, and

he lived long enough to see it opened for traffic. He died in January 1933.

Before moving into the LMS era, consideration should be given to some smaller engines of this period. During the First World War, traffic at Belfast docks expanded so much that little No 42, the dock shunter, was overwhelmed. Malcolm designed a virtually identical replacement 0-4-0ST numbered 16, but with a cab and a steam brake, and this engine acquired the nickname 'The Donkey', possibly because of its hard work at low speed. Over 37 years of service she ran 705,696 miles, calculated on a 5mph average, and was shopped every three years.

The nature of No 16's work brought her into the public view as she crept about her business round the docks. The nominal load for the quays was 12 to 24 wagons, but on one occasion during the Second World War the long-suffering No 16 had the colossal load of 57 wagons. Despite the gloomy prophecies of those who did not know about her high tractive effort and 30 tons adhesion, No 16 confounded them all by getting the train moving. All went well until they crossed the Dufferin Road. At this point a wagon drawbar was pulled out and the train came to a halt across the road, blocking the way for a large number of American GIs making for their ship. A helpful Harbour Policeman reassured them that the blockage would not last for long,

In 1914 traffic must have been brisk because the NCC decided to build a second docks shunter – a carbon copy of No 42 of 1875. This was numbered 16, replacing one of the old H class 2-4-0s. Her works plate bore the legend "MR NCC MAKERS BELFAST 1914". She is seen here at York Road yard. *Author's collection*

In this view, N class No 16 (known as 'the donkey') is working down on the quays in 1935 outside the Bristol sheds. The 'milking stool' on the tank was used to rest the water bag on while filling the tank. The dented cylinder cover shows the dangers of working down at the docks. *Author's collection*

as wagons were usually only moved in tens. As minutes turned to hours, though, he was forced to admit: "They seem to be moving them in hundred and tens tonight"!

No 16 again came to American notice when she was sitting against a rake of coaches in Platform 5 at York Road. Platform 4 was used for loading troop trains, and the men marched in from the docks through the taxi cab entrance beside the Midland Hotel. Bemused by her diminutive size, one GI approached the driver, "Say,

An interesting view of No 16 on shed at York Road in the late 1930s with C1 class 2-4-0 No 57 alongside. *Real Photographs Ltd*

what does this little engine do?". Detecting a note of ridicule in his voice, the driver replied: "Ah, she's kept around here to warm the shaving water!"

To complete our survey of the Midland period, reference must be made to the two steam rail motors built at Derby on Bowman Malcolm's instructions to order No 0 2915, using J1 class boilers with a shortened barrel. Quite unlike anything ever seen before or since at York Road, they are shown in early photographs with the numbers 90 and 91, the oval plates appearing on the carriage portions rather than the engines. This is what may have led Rusk, in his excellent *British Steam Railcars*, to suggest that they were numbered in the carriage series. Later photographs, however, show the number plates in the conventional position on the engines, but with a blank front buffer beam. The coach

underframe was constructed entirely of steel and pivoted to the rear of the engine. The coaches were gas lit and, unusually, had accommodation for first, second and third class, rather than the more typical first and third class accommodation found in other rail motors. When the unit required shopping, the engine could be parted from the coach and driven to the works as an 0-2-2 tank.

Designated the T class, Nos 90 and 91 were delivered in June 1905 and lasted only until 1913. They were designed for short distance runs, but the traffic department used them from Belfast to Ballymena, too long a journey at too fast a speed for such vehicles. To make even this modest journey required a water stop at Greenisland in each direction, and coaling at Ballymena! They had excessive weight on the driving axles and were constantly running hot boxes and big ends. Since the units, with a wheelbase of 49'3½", were far too long for any NCC turntable of the time, the coach end had a driver's compartment. The dimensions of these interesting vehicles are in Table 19.

Articulated rail motors of this type were notoriously prone to vibration and oscillation. The thrust of the connecting rods on a short-wheelbase 0-2-2 tank transmits violent oscillation throughout the engine, and in the absence of the damping effect of buffers, to the carriage as well. As speed built up, riding in the carriage would have become quite disturbingly uncomfortable.

There was a further basic problem for drivers. As the engines were underpowered, even gradients as modest as the bank up to Greenisland slowed them down markedly. A full load of passengers could add three tons to the 13½ tons of the coach – an increase of 20% to the tare weight, and these little engines had no power in reserve. They could have been made bigger, but this would have defeated the purpose of their existence, which was economical working. The fireman had his own problem. With tubes only 4'6" long, opening the firebox door to fire these engines

caused much of the fire to be drawn up and out of the chimney. For this reason, firing was as far as possible done in stations while the engines were stopped, or when they were running with steam shut off.

Nos 90 and 91 were the first engines on the NCC to have outside Walschaerts valve gear – a universal type of motion for Irish rail motors. Another readily observable feature was the external steam pipe which emerged from the dome, then divided and passed down each side of the boiler and under the running plate to the cylinders. A final peculiar feature was their buffers - a long, tapered shank with a head parallel top and bottom but rounded at the sides.

The rail motor scheme was abandoned in 1913 and the engines were scrapped. The coach portions survived, and with new ends and bogies became Nos 79 and 80 of class E1 in the carriage stock – known latterly as 'halt coaches'. As the guard was accommodated in the former driver's compartment, they became brake tri-composites with accommodation for six first, sixteen second and forty-nine third class passengers. Coach 80 could be found running on steam trains well into the era of the Ulster Transport Authority.

Like other railways, the NCC was on the lookout for cheap transport and the combined engine and carriage appeared to offer this. Class T railmotor No 91 is seen here on a bridge near Muckamore, outside Antrim, in 1907 – the first NCC engine to have outside Walschaerts gear. No 91 was in fact an 0-2-2 attached to a coach.

Official NCC photograph

Railmotor No 90 runs into Belfast with the 8.35am Ballymena to Belfast on 16 September 1909. Note the flat-topped buffers and the large outside steam pipe over the top of the boiler. The run from Ballymena was rather long for the little engines, which could be driven from the coach end.

Ken Nunn Collection, LCGB

Table 18: Chronology of non-compound locomotives built 1905–23

Class	Type	No	Deliv	Builder	Cost	Rebuilt	Renumbered	Mileage	Scrapped
T	0-2-2T	90	6/1905	Derby					7/1913
T	0-2-2T	91	6/1905	Derby					7/1913
N	0-4-0ST	16	4/1914	York Rd				705,696	9/1951
U	4-4-0	69	7/1914	Derby	£2996	3/1927 (U2)	2/1923 (71)		1/1956
U	4-4-0	70	7/1914	Derby	£2996	11/1924 (U2)			1/1956
U	4-4-0	14	12/1922	Derby	£6346	2/1937 (U2)	3/1923 (72)		12/1961
U	4-4-0	15	12/1922	Derby	£6346	12/1937 (U2)	3/1923 (73)		6/1956
V	0-6-0	71	2/1923	Derby	£5608	2/1953 (V1) boiler ex-69	4/1923 (13)		8/1964
V	0-6-0	72	2/1923	Derby	£5608	10/1951 (V1) boiler ex-24	4/1923 (14)		5/1961
V	0-6-0	73	2/1923	Derby	£5608	12/1953 (V1) boiler ex-60	4/1923 (15)		12/1961

Notes: V class Nos 71–73 were almost immediately renumbered X, Y and Z, before receiving 13–15.

V class Nos 13 and 15 were reboilered in 1938 using the boilers from U class Nos 72 and 73.

V1 class No 15 was loaded to the GNRB from 5 January to 4 May 1956.

For the later history of Nos 70–73 as Class U2, see page 99.

Table 19: Dimensions of non-compound locomotives built 1905–23

Class/numbers	T class	U class	N class	V class	V1 class
Type	0-2-2T	4-4-0	0-4-0ST	0-6-0	0-6-0
Cylinders	9"x15"	19"x24"	16"x22"	19"x24"	19"x24"
Piston valves	–	8 inches	–	8 inches	8 inches
Coupled wheels	3' 7½"	6' 0"	4' 0"	5' 2½"	5' 2½"
Bogie/trailing wheels	3' 3"	3' 0"	–	–	–
Wheel base	10' 0" (engine)	6' 6" + 6' 7" + 8' 2"	7' 9"	7' 5" + 7' 10"	7' 5" + 7' 10"
Boiler length	4' 6"	10' 4⅜"	11' 9¾"	10' 4⅜"	10' 4¼"
diameter	3' 8"	4' 4¼"	3' 6"	4' 4¼"	4' 0"
tubes	139 x 1¾"	127 x 1¾"/18 x 5⅛"	118 x 1⅞"	127 x 1¾"/18 x 5⅛"	102 x 1¾"/16 x 5⅛"
Heating surface	313 sq ft	1261.5 sq ft	661 sq ft	1261.5 sq ft	1038 sq ft
Firebox (inner)	2'2" x 3'2"	4' 8¾" x 3' 10⅜"	2' 10" x3' 10⅝"	4' 8¾" x 3' 10⅜"	5'2⅞" x 3'4½"
Grate area	7 sq ft	18¼ sq ft	11⅜ sq ft	18¼ sq ft	17⅝ sq ft
Boiler pressure	160 psi	170 psi	130 psi	170 psi	200 psi
Tractive effort	3,798 lbs	17,388 lbs	12,707 lbs	20,031 lbs	23,566 lbs
Weight	39 tons, 3 cwt	47 tons, 12 cwt	31 tons, 1 cwt	45 tons, 4 cwt	46 tons, 15 cwt
Tender weight	–	28 tons, 17 cwt	–	28 tons, 17 cwt	28 tons, 17 cwt
Water capacity	500 gallons	2090 gallons	600 gallons	2090 gallons	2090 gallons
Coal capacity	11½ cwt	6 tons	15 cwt	6 tons	6 tons

Notes: T class – Weight includes carriage; engine only: 25½ tons. Wheelbase of engine and carriage: 49' 3½".

U class – Some sources give cylinders 18" x 24" with TE 15,606 lbs

Chapter 7
Wind of Change: The LMS

Two major changes occurred in the early 1920s, which were to affect the NCC in different but important ways. The first, already referred to, was Bowman Malcolm's retirement in August 1922 after 52 years with the BNCR and MR(NCC). The second was the absorption of the Midland Railway into the new LMS, which was to have a much greater influence over what became the LMS(NCC).

Malcolm's successor was William Kelly Wallace. Wallace was born in 1883, and had worked under Malcolm since 1906. The new CME had been one of Malcolm's promising young men, but although he brought a breath of fresh air into workshop practice at York Road, he was not as strongly placed to argue for what he wanted from his ultimate lords and masters in Derby. Primarily a civil engineer, he took Hugh P Stewart as his assistant and Mechanical Engineer and Freeman Wills Crofts as his Assistant Civil Engineer. Crofts was to become famous as a writer of detective stories, many with railway backgrounds: *Sir John Magill Investigates* in particular has some fascinating, and highly accurate, detail about NCC train services and locomotive operation. Crofts had been District Civil Engineer at Coleraine, and was a friend of the present author's grandfather.

This blending of civil and mechanical matters on the NCC may cause some surprise, but the NCC was a comparatively small concern, and as the late Harold Houston once remarked, "You had to be able to do a bit of everything." WK Wallace stayed at York Road until 1930, when he moved to England. In 1934 he became Chief Civil Engineer of the LMS, certainly the highest position his profession could offer. Another Ulster contemporary of Malcolm's was the chain-smoking Willie Wood, formerly deputy accountant at York Road, and later, as Sir William V Wood, President of the LMS Executive from 1941–1947. The NCC may not have made its English owners much money, but they were amply repaid by the quality of some the staff they gained.

A review of the locomotive position at the start of Wallace's reign shows that there were four modern superheated passenger engines, three modern superheated 0-6-0s, some obsolete 2-4-0s and 0-6-0s and the remnants of the compound era. Goods traffic was not a major problem. Although the NCC carried cattle and general merchandise, freight haulage was well within the capacity of V class Nos 13–15,

compounds 53 and 54, and ageing engines of classes K and L. Passenger traffic was, however, growing more rapidly and, with the LMS about to introduce thirty modern corridor carriages of LMS design for mainline services in 1924, trains were going to get heavier, and some urgent action was needed to get more powerful locomotives. Ten brand new engines were planned for 1924–25, which we shall come to shortly, but York Road could lend a hand as well. Although outmoded, the Compounds had been strongly built and were considered fit for rebuilding – though it might be appropriate at this point to define what 'rebuilding' actually meant.

The famous EL Ahrons (see page 69) defined a rebuilt engine either as one with a new boiler placed on old frames and wheels, or as one with an existing boiler set on to new frames and wheels. By neither of these definitions could the LMS(NCC)'s first two engines, which embodied only the wheels and motion of the originals, be considered rebuilds. Wallace withdrew B class light compounds Nos 59 and 62, and turned out two new engines from York Road which had standard Derby G7S boilers with Belpaire fireboxes (as used on the Fowler 2P class 4-4-0s). Five of these had been sent over by Derby in 1923 and were the same as those used on 4'8½" gauge engines, without the kind of modification for the wider 5'3" gauge which the Midland had always made. At the outset, the new LMS regime at Derby was stamping its authority on Belfast in an unprecedented way.

The first two of the new engines appeared in May and July 1924, resplendent in Midland Lake lined in black with a yellow border and were designated as the 'U1' class. They had the usual 6'0" driving wheels, but only 18"x24" cylinders – surprisingly small, given that the MR 2P class had 20½"x26" cylinders. Nos 1 and 2, and some of the new carriages, were rushed out of the works in time for the Royal Train on 24 July 1924, which conveyed the Duke and Duchess of York to Londonderry. No 1 was the train engine and No 2 stood pilot. Two more of the type followed in 1926 (No 3) and 1931 (No 4). Chronological details appear in Table 21, and mechanical details in Table 20.

In the early 1930s the NCC began to publicise the tourist appeal of its Irish operations, and the U1 class were given the names of some of the Glens of Antrim. No 1 became *Glenshesk*, No 2 *Glendun*, No 3 *Glenaan* (though at first retaining the *Galgorm Castle* nameplates

Top: U1 class 4-4-0 No 2 at York Road on the royal train which conveyed the Duke and Duchess of York (the future King George VI and Queen Elizabeth) to Londonderry in July 1924. The picture was taken in the passenger yard

No 2 was actually the pilot engine whilst No 1 was the train engine. The train comprised four carriages of the new standard LMS stock, just delivered, but the third vehicle is one of the Holden tour saloons (see page 63), specially rebuilt for the occasion.

Official NCC photograph

Above: NCC numbering could be confusing and naming was not much better. Here U1 class No 3 is seen named *Galgorm Castle*. She reused parts from C class 2-4-0 No 33, including the splasher with the name in the beading rather than mounted on top of it. In 1932 she was renamed *Glenaan* in line with the rest of the class. *Lens of Sutton*

Right: U1 class 4-4-0 No 4 *Glenariff* sitting on the turntable at Portrush on 18 June 1932 carrying the white diamond for a branch line train. Note the curved foot plating and splasher beading which distinguished a 'Glen' from a 'Scotch' engine.

William Robb

from No 33 until December 1932) and No 4 *Glenarriff*. If Nos 1 and 2 could claim to be new engines, Nos 3 and 4 could claim no less, as very little of old C class 2-4-0s Nos 33 and 52 would have survived. The solid build of an NCC 4-4-0 is shown in its weight of nearly 51 tons, compared with the much more powerful GNRI Q class which turned the scales at only 49 tons. With a maximum loading of only 17¼ tons on the driving axle, the NCC had few worries about weight restrictions. The U1 class tender was a similarly solid vehicle, with sides raised by a foot to increase water capacity to 2650 gallons.

U1 class 4-4-0 No 4 *Glenariff*, seen here at Limavady on 22 April 1948, has acquired the suffix 'A' to distinguish her from the new 2-6-4T of the same number. She ended her days on the Limavady branch.
HC Casserley, courtesy RM Casserley

Nos. 1 and 2 were first based at Belfast, but with the appearance of the more powerful U2 class a few weeks later, the U1s were dispersed around the system. By 1938, Nos 1, 2 and 3 were shedded at Ballymena, and doing local goods work on the Cookstown line. No 4 remained in Belfast as the spare engine for the

Larne services. There remained one express working for the Ballymena trio, with the through coaches to Larne, detached from the 4pm ex-Derry at Ballymena and worked round the back line via Greenisland. Lightly loaded, but smartly timed at 54 minutes for the 44½ miles to Larne Harbour, this train demanded an average of almost 50mph. A sample appears below of the kind of work which Nos 2 and 3 regularly did on this train.

2 July 1936.

Loco: U1 class 4-4-0 No 3 *Glendun*

Load 3 bogies, 110 tons gross

Miles	Station	Time	Speed
00.0	**Ballymena**	**00.00**	
01.6	Ballymena Goods	02.25	*Very slow start*
04.3	Kellswater	07.38	60
06.5	Milepost 27	09.53	55
	Good mimimum speed over the hump		
08.4	Cookstown Jcn	12.02	60
10.5	Milepost 23	13.48	64
11.7	Antrim	14.58	61
13.7	Muckamore	16.55	61
14.9	Dunadry	18.18	54
17.1	Templepatrick	20.27	52
20.8	Doagh	23.51	51
	Good minimum speed on the bank		
21.9	Kingsbog Junction	24.51	60
22.8	Ballyclare Junction	26.40	61
24.0	Mossley	27.50	63
24.9	Monkstown	28.55	41
	Speed restriction at the junction		
28.6	Greenisland	31.50	30
44.3	**Larne Harbour**	**51.30**	

Note: I am most grateful to my old friend, the late Drew Donaldson, for the use of many of his pre-war records.

Unfortunately no details exist of what must have been a very good run on the Larne line!

No 3 was the first of the 'Glens' to go, when her 1926 firebox wore out in 1946, with a mileage of about 500,000. Latterly, she had been confined to shunting at York Road and working occasionally on the Larne line. The other three engines acquired different boilers,

No 1 receiving 76's old boiler in 1938, and No 2 getting 83's boiler in 1941. No 1 ended her days shunting at Belfast, with a mileage of 645,922 on withdrawal, while No 2 went in 1947 with almost 600,000 miles. No 4 lasted longest, receiving a reconditioned boiler from No 70 in 1945 which kept her going until 1949. She had a varied career, working first from Belfast on main line turns, before being displaced to the Larne line and finally to Derry by the larger U2's. By the end of her life she was numbered 4A to make room for the new 2-6-4 tank which remains in main line service to this

day with the RPSI. Final mileage was 575,417.

Some samples of 'Glen' work on the main line and Larne line will demonstrate the calibre of these engines.

Up 'Portrush Flyer', 1936
U1 No 4 *Glenariff*
Six bogies (170/185 tons)

Miles	Station	Time	Speed
00.0	**Coleraine**	**00.00**	
04.6	Macfin	07.32	63 ½
08.3	**Ballymoney**	**11.35**	
		00.00	
11.7	Milepost 50	07.08	40 ½
15.6	Dunloy	12.12	P.W.S.
18.3	Killagan	15.32	P.W.S.
25.3	Cullybackey	22.49	67
28.2	**Ballymena**	**26.13**	
		00.00	
04.3	Kellswater	06.02	60
08.4	Cookstown Jcn	12.13	PWS
11.7	Antrim	15.25	63
14.9	Dunadry	19.05	
17.1	Templepatrick	21.26	
20.8	Doagh	25.22	44
21.9	Kingsbog Junction	27.53	PWS
7.7	Whitehouse	35.00	67
31.0	**Belfast**	**38.46**	

Date unknown

Train and date unknown
U1 No 4 *Glenariff*
Seven bogies + van (213/225 tons)

Station	Time	Speed
Whitehead	**00.00**	
Kilroot	05.16	44 ½
Eden	05.51	52 ½
Downshire Park	06.30	52 ½
Carrickfergus	**07.54**	
Clipperstown	01.39	
Trooperslane	04.34	31 ¾
Greenisland	06.43	35
Jordanstown	08.28	53 ½
Bleach Green	08.59	54¾
Whiteabbey	09.37	59/60
Whitehouse	10.42	65½/67
Belfast	**14.55**	

The above runs are of high quality in relation to loads, particularly the 1936 'Portrush Flyer' performance, which is possibly the finest surviving run by a 'Glen'. The U1 class was rather harshly treated by the NCC, being placed in the same load classification as the more powerful U2 class with their 19" cylinders.

Contemporary with the U1 engines, Wallace produced a further 4-4-0 class, the well-known 'Castles' of class U2. As York Road was busy with the rebuilds, and Derby preoccupied with LMS design work, the order for the first five went to the North British Locomotive Company in Glasgow and they were delivered in July 1924. For this reason, and although not all the U2s were North British engines, the class became known among railwaymen as the 'Scotch engines'. Since the original U class had by this time been numbered 70–73, the North British engines became Nos 74–78, with maker's numbers 23096–23100, and cost the Company £5328 each.

When naming commenced in 1930, No 74 was named *Dunluce Castle*, 75 became *Antrim Castle*, 76 *Olderfleet Castle*, and 78 *Chichester Castle*, the dates being recorded in Table 22. *Ballygally Castle* nameplates were cast for No 77, but the story went that following Stewart's retirement they were never applied to the engine. However, it is significant that no 4-4-0s were named after June 1933, the month before the first mogul arrived, and it may have been that the naming policy was simply switched to the new larger engines.

So happy were the Board with these engines, that five further engines were ordered in 1925. Nos 79, 80 and 81 were turned out by York Road in August, November, and December 1925 respectively, at a cost of £4813. Nos 82 and 83 were ordered from North British (makers numbers 23171 and 23172) and arrived in May 1925, at a cost of £4920 each. Each of these engines was eventually named: No 79 became *Kenbaan Castle*, No 80 *Dunseverick Castle*, No 81 *Carrickfergus Castle*,

No 82 *Dunananie Castle* and No 83 *Cara Castle*. NCC folklore claimed that No 81 was so named because one of her regular drivers, John Young, was a Carrickfergus man! The ten new engines were augmented by the ongoing programme of rebuilding which supplied eight more engines, basically four rebuilds from the U class in 1924 (70), 1927 (71) and 1937 (72 and 73), and four based on 'heavy compounds' in 1929 (84), 1934 (85), 1935 (86) and 1936 (87). Table 22 details the subsequent history of the whole class and dimensions appear in Table 20.

By 1938, the 'Scotch' engines were the most numerous class on the NCC, and their number was not equalled

until the completion of the WT class 2-6-4 tanks in the early days of the UTA era. By Irish standards, 18 was a large class and the only Great Northern class to exceed this were the Glover 4-4-2 tank engines, with 25. Looking further afield, the MGWR had 20 passenger 2-4-0s while, apart from the huge '101 class', which were mixed traffic engines, the GSWR's biggest purely passenger class was the 20-strong '52 class' of 4-4-0s.

Until the appearance of the moguls, the U2 class were the NCC's principal express passenger engines in the early 1930s, and largely based in Belfast for top-link duties with their own crews. Drivers Murphy and McCall had No 74, Campbell and Nixon No 78,

Above: A nice view of U2 class 4-4-0 No 74, the first of the North British built 'Scotch' engines, at Portrush on 10 August 1930. Note the absence of jacks and the more modern Fowler style tender. She is beautifully turned out in LMS crimson with the smokebox straps and ring burnished. The clerestory bogie behind the engine is dining car No 10.

HC Casserley, courtesy RM Casserley

Right: In general appearance, the U2 class resembled the LMS 2P 4-4-0s, and shared the same boiler. No 77 was photographed at Belfast in July 1932. The famous diamond Works plate of the North British Locomotive Co can be seen on the smoke box underneath the handrail.

JAGH Coltas

McAllister and Kealy No 79, and Young and McKenzie No 81.

When the moguls arrived, the U2s were dispersed rather more widely round the system. In 1933, Nos 74 and 81 went to Coleraine where they remained until the war, to be joined later by Nos 77 (latterly the Coleraine spare engine), 85 and 86. Nos 70, 73 and 80 were based at Larne, while Nos 82 and 87 moved to Londonderry where they were joined during the war by Nos 83 and 84. Nos 71 and 75 were sent to Ballymena. This left Nos 72, 75 and 76 in Belfast. Nos 72, with drivers Patterson and White, and 76 with McNally and McCrory, worked mainly on the Larne line. No 75, with drivers McAuley and Shiels, saw more regular use on the main line.

Above: U class 4-4-0 No 70 was rebuilt as a U2 in November 1924, but her frames were not extended and so she had a shorter cab than the rest of the class. She retained the 'breadcart' tender which was made taller to hold 2650 gallons. She is seen at Belfast about 1929.
Author's collection

Left: This picture shows as yet unnamed U2 class No 80 on the 12.05pm Belfast to Portrush, on 20 September 1932, on the old main line near Monkstown, during single line working in connection with the building of the loop line. This stretch of trackbed was abandoned in 1933 when the 'back line' was realigned on a new single track formation running from the distant bridge towards the right, the temporary line in the foreground giving access to this.
William Robb

Below: No 78 *Chichester Castle* leaves Portrush in 1932. To the right can be seen U1 No 4 *Glenariff* on the turntable road. Names were applied to both classes from 1931.
William Robb

The situation, of course, changed quite frequently as engines visited the shops or were moved around the rural sheds to cover motive power shortages. The following non-stop boat train runs on the Larne line give a flavour of what the 'Scotch engines' could do in their heyday.

	Run	1		2		3		4	
	Loco	78		84		74		84	
	Load	145/152tons		235/250 tons		271/300 tons		277/300 tons	
		Down		Down		Down		Up	
00.00	**Belfast**	**00.00**		**00.00**		**00.00**		33.08	
03.3	Whitehouse	05.37	53	05.15	56	05.58	49	28.36	70
04.3	Whiteabbey	06.58		06.40		-		27.39	
04.6	Bleach Green*	07.36	47			07.55	46		
5.2	Jordanstown	08.20	49/47	08.16			42 *minimum*	26.13	60
6.7	Greenisland	09.56	47	10.52	65	10.40		24.54	54
	Trooperslane	11.14	58/66	12.30	66		67	23.21	40
9.3	Carrickfergus	12.47		14.08		13.50	65	20.54	50
11.0	Eden	13.35	66½	14.46			67		
11.6	Kilroot	14.37	62	15.51	67	15.35	60		
14.7	Whitehead*	17.38	62	18.51	57	18.40	57	15.17	51/56
16.6	Ballycarry	19.33	64	20.51	60	20.34		13.04	53
19.8	Magheramorne	22.44	60	23.59	60	23.32		09.18	53
21.6	Glynn	24.39	58	25.51	61	25.20	66	06.53	48
23.3	Larne Town*	26.33		27.33		27.00	30	04.13	40
24.3	**Larne Harbour**	**29.08**		**30.00**		**29.40**		**00.00**	

*There were permanent speed restrictions at Bleach Green (50 mph), the curves between Kilroot and Whitehead (45 mph) and through Larne Town (40 mph). The Boat train was one of the NCC's most prestigious trains, and every effort was made to keep time, even if the boat was late into Larne, or the connecting services from the main line delayed.

Run 1 shows how No 78 handled the usual five coach formation, with a fast start to Whitehouse, a good climb to Greenisland, and a brisk run right up to the restriction around the curve at Larne Town.

Run 2 shows an even faster start, worked very hard out of Belfast with 100 tons more to pull. The time lost on the climb to Greenisland was recovered by determined running down through Carrickfergus and along the seashore after Kilroot.

Run 3 features the hardest work of the three runs, with a very heavy load well handled. 42mph was an excellent minimum speed at Greenisland, and the engine was steamed down through Carrickfergus. Some very fast running between Carrickfergus and Larne brought the train into Larne Harbour dead on time.

Run 4 in the up direction demonstrates just how sharp the 1 in 98 gradient is between Carrickfergus and Trooperslane. No 84 produced a fine performance with this load, and did well to lose only three minutes on the difficult 30 minute timing.

Opposite: A lovely study of U2 class No 75 *Antrim Castle* at Portrush with a six-wheel van next the engine. The cut out at the rear edge of the cab roof can be seen.

Lens of Sutton

Right: No 85 of 1934 in immaculate LMS red. It was said that ICI supplied a special kind of paint and this was tried on No 85. Its lasting qualities may not have been good because no other engine got it. Note the Belfast builder's plate on the smokebox reading "LMS NCC 1934 BELFAST"

Official NCC photograph

Above: No 76 *Olderfleet Castle* sitting on the turntable road at Portrush in June 1932. Note the LMS crest on the cab side and the cut-away roof. *William Robb*

Above: U2 class No 71, before naming, at Belfast in 1931. Note the full cab roof. Only the engines built by North British had the cab roof cut away. *LGRP*

After the arrival of the moguls in the 1930s, Scotch engine runs on the main line became rather more rare. In keeping with the NCC's tendency to regard Belfast–Portrush as the main line, and Coleraine to Londonderry almost as the secondary route, the latter before the war had neither a mogul allocation nor a turntable long enough to turn one. So four-coupled motive power remained the norm in the North West, and the two runs on the next page illustrate the kind of day-to-day work that the U2s were doing on these trains. The down run shows No 74 on the through coaches from Belfast of the 9.45am ex-Belfast, while in the up direction No 77 had a heavier load on the afternoon train from Derry. The NCC's route from Coleraine to Derry along the Bann and Foyle estuaries was virtually level, requiring constant steaming, and providing no respite to pull round the boiler of an engine shy of steam!

	Loco	74		77	
	Load	Five bogies+van (150 tons)		Seven bogies (230 tons)	
	Date	18 June 1938		17 March 1950	
	Train	09.45am ex-Belfast		3.15pm ex-Derry	
00.00	**Coleraine**	**00.00**		**08.05**	
04.3	Post 66	06.35	55		60
05.08	**Castlerock**	**08.30**		**09.45**	
07.2	Downhill	03.05		08.00	62
10.2	Magilligan	06.25		04.53	
13.2	**Bellarena**	09.38	58/62	**07.11**	
16.3	Post 78	15.25	*TSR*		52
18.0	**Limavady Junction**	**17.59**		**11.05**	
19.7	Ballykelly	03.50		08.45	61
20.7	Carrichue	04.45	50	07.37	
24.3	Post 86	08.21			
26.0	**Eglinton**	10.10		**11.32**	
27.3	Post 89				62
28.7	Culmore	12.50	59	06.30	60
29.5	Lisahally				58
31.3	Post 93	15.31	60		
33.6	**Londonderry**	**19.00**		**00.00**	

As already noted, Belfast–Portrush services were dominated by the moguls after the mid 1930s, leaving the 'Scotch' engines with such undistinguished work as the 6.45am Coleraine–Belfast via the Derry Central, and the 9.55am Belfast–Coleraine which served all stations from Dunadry to Ballymoney and conveyed the Coleraine TPO. There were, however, flashes of brilliance. OS Nock, for example, was fortunate enough to have an outstanding run on the North Atlantic Express in 1938 with No 79 *Kenbaan Castle* and a load of 190 tons gross. Two minutes were dropped on the climb to Ballyclare Junction, passed at 30mph, but No 79 ran down the bank like the wind, sustaining

75mph until beyond Antrim, falling only to 62½ over Cookstown Junction hump, and managing 70½ at Kellswater. Inspired by a late restart from Ballymena, driver Jimmy Marks with fireman Billy Hanley passed Cullybackey in the outstanding time of 6'08" at 47mph, then, after the Dunminning curves, worked up to 64 at Glarryford and 66 at Dunloy. His speed of 76½ down Ballyboyland bank is the highest recorded with an engine of this class. The average speeds were also quite remarkable: the 34½ miles from Ballymena to Portrush were run in 37'25", an average of 55¼mph, and the pass to pass average from Cullybackey to Portstewart was 63½mph.

U2 class 4-4-0 No 72 has just arrived at Larne Harbour with the down Boat Express, which is using the famous rake of North Atlantic Express coaches. As an example of intensive rostering at an early date, when the up NAE arrived in Belfast, the Larne engine was waiting to hook on the back and take the through coaches and Belfast train to Larne Harbour with passengers for the Stranraer ferry. No 72 (originally a U class) was one of the unnamed Scotch engines, the others being Nos 70, 73, 77, 85 and 86. However, all but No 85 had been allocated names before 4-4-0 naming was abandoned with the advent of the moguls.

Author's collection

After the war, the advent of the mogul tanks pushed the U2's even further down the rank order, though in 1951 the 9.25am Belfast–Portrush was surprisingly rostered for a Ballymena U2, and was worked through that memorable summer mostly by No 75, supported on occasions by Nos 74, 80 and 87. A very fine run was done nearly three years later, on 19 March 1954, with No 78 on the 6pm Belfast–Ballymena, an important evening commuter train on which driver John Orr produced something really special. The late

Mac Arnold remarked that Orr habitually worked an engine hard, not for any pleasure that swift movement gave him, but from anxiety about time-keeping. Like many drivers, Orr liked to be photographed with his engine, which was how I first struck up a friendship with him. Riding with him on the footplate on that March evening, I was rewarded with a very competent performance on a train whose modest load was exactly right for a fast run.

Date:	**19 March 1954**	
Train:	**6pm Belfast–Ballymena**	
Loco:	**U2 No 78**	
Load:	**Four bogies + van (130 tons)**	

Belfast	**00.00**	
Whitehouse	07.28	44
Temporary speed restriction at Greencastle		
Whiteabbey	08.56	37
Monkstown	12.17	32
Mossley	14.07	34
Good climb of the bank considering TSR at Greencastle		
Ballyclare Jcn	16.20	35
Kingsbog	17.34	39
Doagh	19.35	58/60
Post 15	21.27	62
Templepatrick	22.54	67
Dunadry	24.54	69½
Muckamore	26.18	60
Post 21	27.24	60
Antrim	**28.49**	

Sadly, time was running out for the 'Scotch' engines. In the mid 1950s the UTA had 33 modern six-coupled engines and 18 4-4-0s. The slaughter began in earnest in 1956, when eight Scotch engines were withdrawn, followed by two more in the following year. One odd feature of the first wave of withdrawals was that they were nearly all uneven numbered engines – Nos 70, 71, 73, 75, 77, 79, 82 and 83 in 1956 and Nos 81 and 87 in 1957. The main line and Larne line were largely dieselised in 1958–59 and by early 1960 four more had gone – Nos 76, 78, 85 and 86. At the end of 1961, Nos 72, 80 and 84 followed, though of these only No 80 had been active. The final survivor of the class was No 74 *Dunluce Castle*, and Harold Houston used his influence, not only to have her preserved, but also externally finished in full NCC livery to the highest of standards. Her last journey to date was on 14 February 1993, when she was taken from the museum in Witham Street by road to Adelaide yard and hauled to the Ulster Folk and Transport Museum's new transport gallery at Cultra.

Although No 74 was only hauled dead over a few miles of Great Northern track on this occasion, two of the class were hired by the GNRB for a period between 1952 and 1954. No 81 arrived first, on 1 May 1952, replaced by No 72 on 12 March 1954. The intention

A broadside view of U2 class No 87 *Queen Alexandra* in full red livery at Belfast on 22 June 1937. This engine was constructed in 1936 using parts from A class compound No 63. It is possible the old splashers were used and the name plates left in the same place.
HC Casserley, courtesy RM Casserley

Left: A fine view of Whiteabbey station in August 1933 before colour light signalling was installed. U2 class 4-4-0 No 82 *Dunananie Castle* has a Larne line local train. The tall signals aided sighting on the road bridge.

William Robb

Below: U2 No 86 at Londonderry in 1937. It was said she was to get the name *King Edward VII* from compound No 33 to match No 87 *Queen Alexandra*, but she never carried it.

OS Nock, courtesy CP Friel

had been to send No 73, but she was just ex-works and not sufficiently run in. No 72 stayed on the GNRB until 29 September 1954. These engines worked mostly on Belfast–Clones trains, whose relatively easy schedules did not particularly tax their steaming capacity.

Comparisons are inevitably made between similarly sized engines from different companies, and the Scotch engines might be compared with the GNRI Q class. The NCC engine, weighing in at 51½ tons and with an axle loading of 17¾ tons, was the more solidly constructed of the two: cracked frames were a rare occurrence. By contrast, the Q class weighed 49 tons with an axle load of 16¼ tons. Before the rebuilding of the Boyne Viaduct in 1932, a maximum axle loading of 17 tons was the condition that the GNRI Civil Engineer imposed on the locomotive department. The result of this limitation was a tradition of shallow-framed engines, and a susceptibility to cracked frames.

Boilers in the two engines were of similar size. The U2 had a standard LMS G7S boiler, 10'6" long and 4'6½" in diameter, and we have already noted that the 6'4"x3'4" firebox with 21 square feet of grate was not widened to take advantage of the more generous Irish gauge. In passing, it may be noted that when William Stanier came to the LMS he quickly became aware of the narrowness of the U2 firebox and instructed that, as the Scotch engines went through the works for

overhaul, their fireboxes should be widened. Perhaps he remembered his earliest days, working with the last of the GWR 7'0" gauge engines!

The Q class boiler was 11'0"x4'6", the firebox was 5'10" long and the grate area 20 square feet, the Great Northern taking full advantage of the Irish gauge to build an engine with a wider firebox. This, combined with a longer coupled wheelbase and a flatter grate, made the Great Northern 4-4-0 capable of sustaining effort longer – very necessary for a class of engine which regularly worked over Carrickmore bank on the difficult section between Dungannon and Omagh.

Mechanically, the Q class had 18½" cylinders with a 26" stroke, while the U2s had 19" cylinders but a 24" stroke. Both had 8" piston valves with the short travel customary at the time, and while the Q's has Stephenson

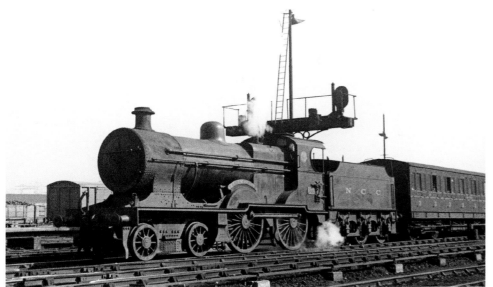

Left: No 83 *Carra Castle* at Belfast on 22 April 1948. Her bypass valves have gone.
HC Casserley, courtesy RM Casserley

Below: No 72 was the first of the 'wandering' Scotch engines and is seen here at Adelaide, on the GNRI, with the morning train for Cavan in 1952. The tablet catcher has been removed – it would have been of no good on the GNRI.
Author

gear, the NCC remained loyal to the Walschaerts system. On the Q class the screw reverser gave a finer adjustment of steam supply than the lever on the U2.

Great Northern men were fond of the Qs because of their free steaming and haulage power. Some even preferred them to the larger S class, maintaining that the flexing of their lighter frames gave a more comfortable ride over the curving Derry road than the bigger engines. On the other hand, Dundalk's fitters found the Qs a sore trial; those lighter frames may have flexed on the Derry Road, but all too often they cracked. George Glover, the Great Northern's CME, stated that no engine should be allowed to work for more than six months at a time on this difficult section. For their part, NCC men swore by the Scotch engines, with their comfortable riding and their economy. They had to be persuaded to accept the much harder-riding moguls when they appeared from 1933.

In summary, the Q class was probably stronger and more able to sustain hard steaming, whereas the Scotch engines were mechanically sounder and easier to maintain. By coincidence, Q class No 131 and Scotch engine No 74 have both survived in preservation. Is it too much to hope that at some time in the future they might once again be pitted against each other?

Above: No 81 *Carrickfergus Castle* is seen here on the GNRI approaching Balmoral on the 1.16 pm SO to Lisburn. The sidings to the right (known as the RUAS sidings) served the Balmoral agricultural show.
A Donaldson

Right: U2 class 4-4-0 No 87 *Queen Alexandra* is pictured in UTA days at Londonderry, on 13 April 1955.

Kelland Collection, Bournemouth Railway Club

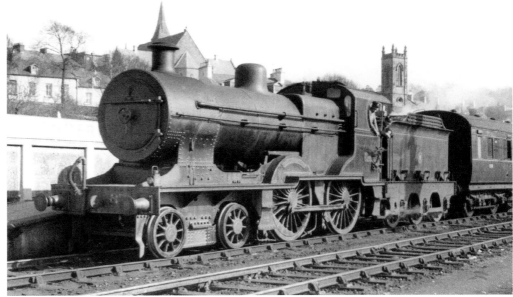

Below: No 78, at a much later stage in her career, crosses the Bleach Green viaduct with a football special in 1958.

A Donaldson

Above: Displaced from the main line, the Castles found work on secondary duties, one of which was a strange Cullybackey–Antrim–Aldergrove working. No 84 is seen on this service at Aldergrove about 1952. *Author*

The A1 class 4-4-0s

From the Scotch engines, we move on to the A1 class. These were rebuilds of the A class 'heavy compounds' of 1901–08. Of the thirteen in this class, No 20 was rebuilt as a U2 in 1929, but the original intention seems to have been to rebuild the rest as A1 because, in the mid-1920s the previously scattered numbers were reorganised to put the whole class in the series 33, 34, 58–69 (though 60, 61 were B1 class). However, in a change of plan, Nos 59, 63 and 67 were rebuilt as U2 in 1934–36, becoming 86, 87 and 85 respectively.

This left Nos 33, 34, 58, 62, 64, 65, 66, 68 and 69 which became Class A1 and received the smaller G6 boiler. These boilers were high-pitched – 8'9" to the boiler centre line – with a lot of daylight between boiler and frames. The chronology of the A1 class appear in Table 24, and the dimensions in Table 20.

It is not easy to see what the NCC expected of these engines which, with only

Left: No 80 *Dunseverick Castle* leaves Antrim on 16 May 1956. Stanier considered the maintenance cost of the Fowler-Anderson bypass valves to outweigh their advantages and had them removed.
Kelland Collection, Bournemouth Railway Club

Below: A1 class 4-4-0 No 68 *Slieve Gallion* at Ballyclare in 1936. This was the first of the class to be rebuilt (in 1927) and retained the small 'bread cart' tender with coal rails. The crew and guard are keen to be photographed with an engine which is superbly turned out.
LGRP

18"x24" cylinders and 4'0" diameter boilers, were the equals neither of the Scotch engines nor the Glens. In the late 1930s they were to be found around the system: Nos 34 and 66 at Coleraine, 33 and 65 in Belfast, 58 and 69 at Larne, and 62 at Cookstown, breaking the previous compound monopoly in this shed. Significantly the 200lb engines (Nos 33, 58 and 69) were sent to sheds most likely to need them.

With most of the Scotch engines and Glens, as well as at least eight of the moguls, available by the mid 1930s, the A1s were given little opportunity for express work. Three examples may be quoted, two of them from Portrush to Coleraine, the first with No 62 and a 92 ton train, and the second with No 65 and 125 tons. The lighter train stopped in Portstewart in 5'50" sustaining 24mph on the climb through Dhu Varren, and the heavier train took 7'25", with a minimum of 19½mph. Both trains ran from Portstewart to Coleraine in six minutes, No 62 whisking her three bogies away to a rare 60mph. No 62 is recorded as having run from Coleraine to Castlerock in 8'30". In the other direction, a respectable run with No 34 and 150 tons shows a time of 8"49" from Castlerock to Coleraine. It is most unfortunate that no details survive of dates, or of the trains these engines were working – particularly that with No 62, which seems to have been working through from Portrush to Derry. More detailed, though still lacking date or precise train, is a fine run on a Larne Boat Train with No 62. This was a quite exceptional effort, with some very high speeds made possible by

the NCC's use of mechanical tablet exchangers on the single-line section beyond Whitehead.

	Loco	62	
	Load	5 bogies+van 163/175tons	
00.00	**Belfast**	**00.00**	
03.3	Whitehouse	06.18	50
04.6	Bleach Green*	07.54	44
6.7	Greenisland	10.28	47
9.3	Carrickfergus	13.52	59
11.6	Kilroot	15.52	60
14.7	Whitehead*	19.11	59
16.6	Ballycarry	21.07	63
19.8	Magheramorne	24.11	64
21.6	Glynn	25.56	62
23.3	Larne Town*	27.43	30
24.3	**Larne Harbour**	**29.57**	

PSR (Permanent speed restriction)

Right: No 34, when rebuilt in April 1928, at first retained its original name *Queen Alexandra*. Some early rebuilds to Class A1 also had their tenders converted to the flat high sided type seen here.

Author's collection

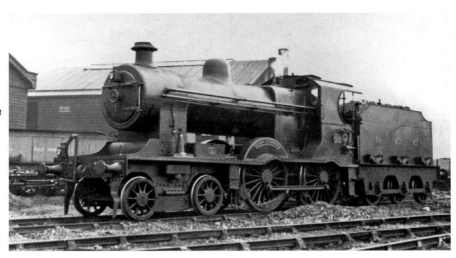

Below: A1 class No 34 after renaming as *Knocklayd* in November 1932. The A1 class were named after mountains in the NCC area. Note the combined footstep and sandbox ahead of the leading splasher.

Author's collection

Above: No 33 *Binevenagh* of 1932 bringing carriages up from Portrush harbour branch in 1938, with plenty of native sand on the rails. Barry's amusements is in the background.

A Donaldson

Left: No 66 *Ben Madigan* at Belfast on 9 August 1930, a few months after rebuilding in May that year. She was a favourite on the Larne boat trains and carries the white diamond on the smokebox indicating a Larne line train.
HC Casserley, courtesy RM Casserley

Right: A1 class 4-4-0 No 65 of 1929 was photographed unnamed at Ballyclare on 9 August 1930. She became *Knockagh* in February 1931. She has the usual branch line rake of two flat-sided bogies.
HC Casserley, courtesy RM Casserley

Below: No 62 *Slemish* with the 1.20pm from Larne Harbour at Whitehead in June 1935. She has a high sided tender with coal rails but the coal looks dubious. The train includes three of the 'Larne steels' of 1933 but attached to a much older brake composite and van.
William Robb

Left: No 58 at Dungiven on 7 July 1949. By this time the turntable had gone and the engine has been fitted with a tender cab. Driver J Faulkner is in the foreground.

No 58 was the only unnamed A1 engine, but before naming of the 4-4-0s was abandoned in 1934 she was to have been *Lurigethan*. The last three engines to be rebuilt (Nos 33, 69 and 58 in that order) received 200psi boilers, giving them a tractive effort of 18,360lbs in contrast to the 14,688lbs of the 160psi machines.

RN Clements

Right: An interesting view of A1 class No 62 at Portrush goods yard on 10 August 1930, two months before being named *Slemish*. The high built-up tender had a capacity of 2650 gallons. The bracket signal, facing Coleraine, indicated your arrival platform and was replaced, in the 1932 resignalling, by a single arm with illuminated indicator box. Note the stacked boxes of permanent way bolts and the long ladder used by signalmen to place and refill the oil lamps.

HC Casserley,
courtesy RM Casserley

Left: No 69 *Slieve Bane* at Belfast on 14 July 1933. The diamond and headcode suggest she is about to work the down boat train. Note the low tender in contrast to that on No 62 above. *LGRP*

Opposite right: B3 class 4-4-0 No 24 is seen about 1936, running tender-first with a single coach Greenisland local, possibly substituting for a railcar. The carriage is a conversion from one of the 1905 railmotors (see page 76). No 24 was named in 1932 and withdrawn in 1947.

Real Photographs Ltd

Left: No 62 on a local at Ballymena in 1946. The leading coach is one of the J10 open thirds built in 1937–39. These were converted to railcars 6–13 in 1951–53 and rarely appear in photographs as hauled stock.

Kelland Collection, Bournemouth Railway Club

Below: No 64 *Trostan* at Coleraine, where she was a favourite. Note the 1929 works plate on the sandbox. She spent the last ten years of her life working between Coleraine and Londonderry and on the Limavady branch.

RG Jarvis, Midland Railway Trust, Ltd

The B3 class 4-4-0s

If the A1s were an unusual rebuild, the B3s, or 'Whippets', must count as a positively grotesque one. I have already described how the U1 class were constructed using parts from B and C class compounds, becoming virtually new engines. This left B1 class Nos 60 and 61, B2 class No 24 and C class 2-4-0 Nos 21, 51, 56, 57 and 58. Of these, Nos 21, 56 and 57 were not rebuilt, other than the fitting of 'five foot' boilers to Nos 21 (renumbered 51) and 57. The other five were rebuilt to Class B3, as detailed in Table 23.

It was in the rebuilding to Class B3 that these engines took on their singular appearance. Unlike the U1 class, which used parts from the B and C classes but had new frames with a longer 8'2" wheelbase, the B3 class kept their old frames (with a rear extension)

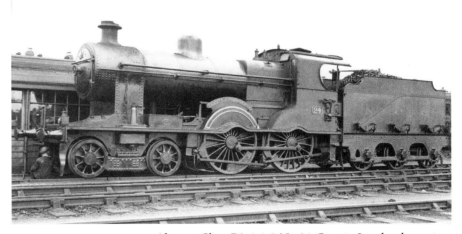

Above: Class B3 4-4-0 No 24 *County Londonderry* at Belfast 1n 1936. The box-shaped projection below the running plate above the bogie is a Fowler Anderson by-pass valve. The driving wheels show clearly the type of 'bolt on' balance weights favoured by the NCC – large on the front drivers and small on the rear. No 24 was rebuilt in 1928 from the unsuccessful saturated B2 rebuild of 1925 (see pages 59 and 60). The early 'Whippets' had 160psi G6S boilers and their tenders were rebuilt with tall straight sides.

LGRP

Left: B3 class No 61 *County Antrim* at Londonderry on 15 May 1937. Nos 60 and 61 were rebuilt from two B1 class compounds in 1932 and were given 200psi G6AS boilers. The tenders were not rebuilt and the engines were named from the start. The jacks mounted beside the smokebox were carried by all older NCC engines. These views show clearly the high pitched boiler which led to the nickname 'Whippet'. The high pitch was necessary to keep the firebox and grate clear of the trailing axle, as the 7'0" coupled wheelbase gave insufficient room to drop the firebox between the axles. In contrast the 8'2" coupled wheelbase of the A1 class allowed a lower pitched boiler.

RG Jarvis, Midland Railway Trust, Ltd

Right: No 21 *County Down* at Belfast on 20 March 1939. This engine has lost its Fowler Anderson by-pass valve. As with most of the rebuilds, the NCC works plate is mounted on the sandbox. No 21 was rebuilt from a C class 2-4-0 in 1928, named in 1932 and was scrapped in 1947.

RG Jarvis, Midland Railway Trust, Ltd

Right: The NCC goods traffic was rarely heavy and usually well within the compass of a four-coupled engine. No 61 has a Greenisland to Larne goods in this view at Whitehead in 1934 but is incorrectly carrying the lamp code for a local passenger train. She should also have a lamp on the left side of the buffer beam.

Real Photographs Ltd

Left: B3 class 4-4-0 No 28, photographed some time between rebuilding in June 1927 and February 1932, when it was named *County Tyrone*. This engine was converted from C class 2-4-0 No 58 and was withdrawn in December 1938, long before the remaining four.

RK Blencowe, Derek Young collection

Left: The 'Whippets' did considerable work on the Larne line – the lighter trains suited them. On 14 May 1933, No 60 *County Donegal* is steaming hard at Jordanstown with an up train. The path beside the engine ran from Bleach Green to beyond Greenisland. It was laid by the NCC to enable those who lived beside the line to walk easily to their nearest station. The railway of the 1930s did everything to encourage passengers to use trains. Today's 'Auschwitz style' fencing was 70 years in the future.

W Robb

Right: No 60 *County Donegal*, prepares to depart from York Road Platform 1 in the late 1930s. The overall roof in the background was destroyed during the Blitz in April 1941.

Alan Whitehead, courtesy CP Friel

and the original 7'0" wheelbase. However, since two of them were rebuilt from 2-4-0s, it is possible that they used the old B class 4-4-0 frames from Nos 59 and 62, withdrawn in 1924 to create the first two U1s.

The original boiler centre-line of the B and C class engines had been 7'7" from the rails, and the firebox had a small flat grate between the driving axles. The rebuilds had the G6S boiler, which had to be pitched with a centre line no less than 8'9" from the rail to accommodate the much larger firebox, which had a grate sloping up over the rear axle. The cab was perched on the extended frames, giving the engine its strange appearance, and the class its nickname of 'Whippets' – a smaller version of a greyhound!

Of the five B3 class, Nos 24, 60 and 61 retained their original numbers, but the former 2-4-0s, Nos 51 and 58 were renumbered 21 and 28, in what may have been a partially implemented plan to number the whole five in the 20 series, or perhaps just to distinguish them from the remaining C class 2-4-0s. In 1932 all five 'Whippets' were named after counties.

Although the nickname spread to the A1s, it was strictly speaking coined for the B3s. The earliest withdrawal was No 28 *County Tyrone* in 1938. Of the engines which survived the war, Nos 21 and 24 were usually found in the Belfast area. No 21 was something of a Houdini; withdrawn in July 1936, reprieved four months later, withdrawn again in September 1940 and again reinstated in July 1941. Nos 60 and 61 were rarely seen in Belfast, but frequented the Coleraine–Derry section and the Derry Central line. These two engines had boilers pressed to 200 lbs.

The dimensions of the class can be found in Table 20. Performance details of the B3 class are scanty, but three runs survive between Castlerock and Coleraine. No 61

County Antrim had three bogies, three four-wheeled vans and a bogie van, gross weight 150 tons, and ran from Castlerock to Coleraine in 8'14" with a maximum of 55mph. The same engine with three bogies, a bogie van and two four-wheeled vans covered the same section in 7'51" with the excellent maximum speed of 58½mph. Finally, No 21 *County Down* with seven compartment bogies, ran the section in 8'10", with a maximum of 55mph. In view of the interesting train compositions, it is doubly unfortunate that no details survive of either the trains these engines were working, or the dates on which they worked them.

Even more interesting is the discovery of a 'Whippet' log on the Derry Central, which produced some fine work with a light train and the advantage of a 200lb boiler. The engine was No 60 *County Donegal*, and the load four bogies, or 90 tons. The 2.2 miles from Cookstown Junction to Randalstown were run in 3'51', maximum 43mph, and thence to Staffordstown (5.4 miles) in 8'07" with an excellent 58mph. The 3.9 miles from Staffordstown to Toome were run in 5'52" with a best speed of 55, and 52mph was achieved between Toome and Castledawson (4.3 miles). The final 2.3 miles to Magherafelt were run in 6'01", with a maximum of 37mph.

Two further engines come within the scope of this chapter, neither of them ever noted for speeds such as have been quoted above! In Chapter 6 reference was made to the two railmotors delivered during Malcolm's regime. A further experiment along these lines was the brain child of WK Wallace – a Sentinel steam railcar

which came in May 1925, followed a month later by a locomotive. The railcar took the number 401 in the carriage series, while the locomotive was numbered 91. Both engines were of similar, and of course diminutive, dimensions, weighing but 20 tons. The cylinders were 6¾" x 9", and the vertical boiler stood 4'4½" tall and was 2'8½" in diameter, pressed to 275lbs. The driving wheels were 2'6" diameter, and the wheelbase 8'6". Fuel capacity was modest – each engine could carry 300 gallons of water and 13cwt of coal. Quoting these dimensions is enough to demonstrate that, from the outset, the Sentinels' boilers were too small to make enough steam, and the driving chains stretched and broke. Both were scrapped in 1932.

In fairness, it must be said that, for the LMS, these were experimental vehicles, and the company learnt a lot from their deficiencies. They went on to introduce a fleet of 14 modified Sentinel railcars on the main LMS system in 1926–27 and six Sentinel shunting locomotives in the early 1930s, all of which gave good service. Sentinels were also widely used on the LNER and in industry. As late as the 1970s, RB Tennant of Coatbridge were happily using a fleet of Sentinels to shunt at their Whifflet Foundry, where the author had the pleasant experience of driving one of these excellent (and beautifully maintained) little locomotives.

To summarise a complex period of locomotive history, the early LMS years saw the NCC put into stock four new U1 class engines and ten new U2s, as well as carrying through an extensive rebuilding programme. At the start of LMS ownership in 1923

Sentinel 0-4-0T No 91 is seen here with her train of 3rd class six-wheeler No 162 and ex-railmotor bogie No 79. Even this was too much for the small boiler when the NCC gradients were encountered and the engine was scrapped in 1932.

Official NCC photograph

there were 29 compounds: seven C class 2-4-0s; five B class 4-4-0s; 13 A class 4-4-0s; two E class 0-6-0s and two D class 4-4-0s. By 1934 there were but three C class, one A class, two E class and a single D class. By 1933 the B class had all been rebuilt. Of the Cs, only Nos 51 (ex-21), 56 and 57 remained. Of the most numerous type, the A class, only No 63 remained until 1936 before rebuilding. The two goods engines

were unaltered, but of the 7'0" D class, No 50 became a simple in 1926, and only No 55 remained to be the last broad gauge compound. In this, she outlasted No 54 by only a few weeks. Though hardly a slaughter on the scale of the LNWR after Webb, this was a period of radical change on the NCC. The next decade was to see changes, before which even this period paled into insignificance.

In addition to No 91, the NCC also purchased a Sentinel railcar, numbered No 401 in the carriage list. She is seen here at Kilroot when the line was still single beyond Carrickfergus. Kilroot's handsome station buildings and cabin are now gone and little trace of the station remains.

Official NCC photograph

Sentinel railcar No 401 and locomotive No 91 together at York Road shed on 5 August 1930. The railcar suffered regularly from broken driving chains. The NCC pair were an early experimental design and much less successful than those introduced on the parent LMS system after 1926.

HC Casserley, courtesy RM Casserley

Table 20: Dimensions of locomotives built or rebuilt in early LMS days 1924–32

Class/numbers	U1 class	U2 class	B3 class	A1 class	Sentinel
Type	4-4-0	4-4-0	4-4-0	4-4-0	0-4-0T
Cylinders	18"x24"	19"x24"	18"x24"	18"x24"	6¾"x9"
Piston valves	8 inches	8 inches	8 inches	8 inches	
Coupled wheels	6' 0"	6' 0"	6' 0"	6' 0"	2' 6"
Bogie wheels	3' 0"	3' 0"	3' 0"	3' 0"	
Wheel base	6' 6" + 6' 6" + 8' 2"	6' 6" + 6' 7" + 8' 2"	6' 6" + 6' 6" + 7' 0"	6' 6" + 6' 7" + 8' 2"	8' 8"
Boiler length	10' 5⅞"	10' 5⅞"	10' 5⅞"	10' 5⅞"	2' 8½"
diameter	4' 6½"	4' 6½"	4' 0"	4' 0"	4' 4½"
tubes	148 x 1¾"	148 x 1¾"	102 x 1⅞"	102 x 1¾"	
Superheater tubes	21 x 5⅛"	21 x 5⅛"	16 x 5⅛"	16 x 5⅛"	
Heating surface	1421.3 sq ft	1421.3 sq ft	1038 sq ft	1038 sq ft	
Firebox	7' 0"	7' 0"	5' 10⅞"	7' 0"	
Grate area	21.1 sq ft	21.1 sq ft	17⅝ sq ft	17⅝ sq ft	
Boiler pressure	170 psi	170 psi	160 psi	160 psi	275 psi
Tractive effort	15,606 lbs	17,388 lbs	14,688 lbs	14,688 lbs	3195 lbs
Weight	50 tons, 14 cwt	51 tons, 10 cwt	45 tons, 2 cwt	46 tons, 7 cwt	20 tons
Tender weight	32 tons, 2 cwt	32 tons, 19 cwt	26 tons, 18 cwt	28 tons, 17 cwt	
Water capacity	2690 gallons	2500 gallons	2120 gallons	2090 gallons	300 gallons
Coal capacity	6 tons	5 tons	6 tons	6 tons	13 cwt

Notes: U2 class Nos 70–73, 84–87 had 2690 gallon tenders, as on Class U1.

U2 class Nos 72, 73 and 86 received 2500 gallon tenders in 1954.

U2 class Nos 72 and 81 were given 18" x 24" cylinders in 1945 and 1946 respectively (TE 15,606 lbs).

U2 class Nos 72, 73, 85, 86 and 87 had wider fireboxes with a 23 sq ft grate and 123/18 tubes, HS 1292 sq ft.

B3 class Nos 60, 61 and A1 class Nos 33, 58 and 69 had 200 psi boilers and a TE of 18,360 lbs.

A1 class Nos 62 and 66 had 2690 gallon tenders.

B3 class No 60 had a 2690 gallon tender and No 61 a 2090 gallon tender.

Table 21: Chronology of the U1 class 4-4-0s

No	Name	Deliv	Builder	Cost	Parts from	Boiler change	Mileage	Scrapped
1	Glenshesk (6/1932)	5/1924	York Rd	£3601	B 4-4-0 No 59	1938 (ex-76)	645,922	4/1947
2	Glendun (5/1932)	7/1924	York Rd	£3509	B 4-4-0 No 62	1941 (ex-83)		4/1947
3	Glenaan (12/1932)	11/1926	York Rd	£3741	C 2-4-0 No 33			10/1946
4	Glenariff (1/1931)	1/1931	York Rd	£3032	C 2-4-0 No 52	12/1945 (ex-70)	575,417	4/1949

Notes: Nos 1 and 2 were the first engines to receive the LMS crimson lake livery and the first with Belpaire fireboxes.

When built, No 3 carried the *Galgorm Castle* nameplate from No 33 until August 1931, at which point it was transferred to No 52.

No 4 was renumbered 4A in May 1947 when WT 2-6-4T No 4 was delivered.

Table 22: Chronology of the U2 class 4-4-0s

No	Name	Deliv	Builder	Rebuilt from	Boiler change	Mileage	Last used	Scrapped
70		11/1924	(York Rd)	U 4-4-0 No 70	1945?	1,177,466	1951	1/1956
71	Glenarm Castle (6/1932)	3/1927	(York Rd)	U 4-4-0 No 71	1945 (ex-84)		8/1950	1/1956
72		2/1937	(York Rd)	U 4-4-0 No 72	12/1953 (ex-87)		1959	12/1961
73		12/1937	(York Rd)	U 4-4-0 No 73		1,028,954	1955	6/1956
74	Dunluce Castle (7/1931)	7/1924	NBL 23096			1,135,484	1961	4/1963
75	Antrim Castle (11/1931)	7/1924	NBL 23097		3/1953 (ex-2)		1955	6/1956
76	Olderfleet Castle (4/1932)	7/1924	NBL 23098		1938 (ex-78) 10/1953 (ex-3)	1,039,364	1956	9/1959
77		7/1924	NBL 23099			1,025,497	1953	1/1956
78	Chichester Castle (1/1932)	7/1924	NBL 23100		10/1936 (new) 5/1953 (ex-86)	1,069,364	1958	3/1960
79	Kenbaan Castle (7/1931)	8/1925	York Rd				1954	1/1956
80	Dunseverick Castle (6/1933)	11/1925	York Rd		11/1955 (ex-72)		11/1961	12/1961
81	Carrickfergus Castle (12/1930)	12/1925	York Rd				8/1957	7/1957
82	Dunananie Castle (5/1932)	5/1925	NBL 23171		1939 (ex-1)		1952	1/1956
83	Carra Castle (5/1932)	5/1925	NBL 23172		1941 (ex-??)		1949	1/1956
84	Lisanoure Castle (9/1931)	12/1929	York Rd	A 4-4-0 No 20	1945?		1960	12/1961
85		5/1934	York Rd	A 4-4-0 No 67	1949?		1957	3/1960
86		1/1935	York Rd	A 4-4-0 No 59	8/1949 (ex-85)		1957	3/1960
87	Queen Alexandra	5/1936	York Rd	A 4-4-0 No 63	12/1949 (ex-86)	589,090	11/1956	8/1957

Notes: The following names were allocated but not applied: No 70 *Portmuck Castle*; No 72 *Shane's Castle*; No 73 *Carn Castle*; No 77 *Ballygally Castle*; No 86 *King Edward VII*. No name was allocated to No 85.

No 70 did not have its frames lengthened, on rebuilding, and had a shorter cab than the others.

Nos 72, 73 and 86 were given Fowler 2500 gallon tenders (ex-83, 77 and 82 respectively) in 1954.

No 74 *Dunluce Castle* is preserved at the Ulster Folk and Transport Museum, Cultra.

No 78's new boiler in 1936 had the wider firebox (see opposite), likewise transfers from engines with the wider firebox.

Table 23: Chronology of the B3 class 4-4-0s, all built at York Road

No	Name	Deliv	Parts from	Boiler change	Mileage	Last used	Scrapped
21	County Down (7/1932)	12/1928	B1 4-4-0 No 51			1946	6/1947
24	County Londonderry (10/1932)	9/1928	B2 4-4-0 No 24	1945 (ex-58)	545,876	1947	6/1947
28	County Tyrone (2/1932)	6/1927	C 2-4-0 No 58		285,421	1938	12/1938
60	County Donegal (6/1932)	6/1932	B1 2-4-0 No 60			1946	10/1946
61	County Antrim (1/1932)	1/1932	B1 2-4-0 No 61			1946	10/1946

Notes: No 21 was withdrawn in July 1936 and reinstated in November 1936. it was also withdrawn in September 1940 and reinstated in July 1941.

Table 24: Chronology of the A1 class 4-4-0s, all rebuilt at York Road from A class 4-4-0s

No	Name	Deliv	Boiler change	Mileage	Last used	Scrapped
33	Binevenagh (12/1932)	12/1932			1949	11/1949
34	Knocklayd (11/1932)	4/1928		660,278	9/1949	10/1950
58		2/1934	1945	642,370	8/1952	8/1954
62	Slemish (10/1930)	7/1928			1950	8/1954
64	Trostan (2/1932)	8/1929			1950	8/1954
65	Knockagh (2/1931)	6/1929			1949	10/1950
66	Ben Madigan (5/1930)	5/1930			1950	8/1954
68	Slieve Gallion (3/1932)	12/1927	1939 (new)		1947	11/1947
69	Slieve Bane (6/1933)	6/1933			1951	8/1954

Notes: No 58 was allocated the name *Lurigethan* but it was not applied as naming of 4-4-0s ceased after 1933. It later received a tender cab for working on the Limavady branch.

The following names were allocated, but not used, as the locomotives were rebuilt to the U2 class:
No 59 *Craiggore*;   No 63 *Ben Bradagh*;   No 67 *Slieveannora*;

Chapter 8
Malcolm Speir and the Moguls

The NCC entered the 1930s having lost two of its senior officers. William Wallace moved to the LMS in September 1930, and James Pepper, Manager and Secretary, resigned due to ill health at the end of December. Wallace was succeeded as Chief Mechanical and Civil Engineer by Hugh Stewart. Of an old railway family, Stewart joined the BNCR in 1876 as Malcolm's apprentice, before moving to Harland & Wolff's shipyard and finally going to sea as an engineer officer on steam ships plying from Liverpool. This was an unusual, but by no means unique, career path for a railway engineer – Charles Collett of the GWR had, for example, worked for Maudsleys, the marine engineers – and eventually Stewart returned to the NCC as Works Manager and finally assistant to Wallace.

The new Manager and Secretary was the legendary Malcolm Scott Speir. Beginning in the traffic department of the Midland Railway, he became a Headquarters Inspector at Derby and, after a period studying railway operation in the USA, returned to the Caledonian Railway as outdoor Assistant General Superintendent. He served with the Royal Engineers in World War One, winning the Military Cross and attaining the rank of Major. This title not only identified him, but also distinguished him from his brother, Colonel Speir, who also served the LMS, but at London Euston.

After the war, Malcolm Speir returned to the Caledonian and, when the LMS was formed, he became General Superintendent of the Northern Division before taking command at Belfast. A bachelor and something of a workaholic, Speir lived in the Midland Hotel and expected his orders to be carried out promptly and accurately. His knowledge of railway operation was as immense as it was detailed, though relations with his heads of departments were often strained as he interfered constantly in the running of every department of the NCC, ruthlessly sidelining those who had the temerity to challenge him.

Positively, he was a great railwayman who fought to uphold railway interests. He realised the importance of tourist traffic and the network of railway hotels. He introduced a range of unlimited travel weekly rover tickets, aimed not just at tourists but at readers of the *Railway Magazine* "whose interest is in the timing of locomotive performance". He had a vision of road-rail container traffic which was far ahead of its time, and believed the railway should diversify into both bus and road freight operation. As air transport began, he believed that railways should be running their own airlines as well as shipping services.

Though not loved, Speir was deeply respected and, in the bleak years of railway decline under the UTA in the 1950s, men of all departments would often wish that 'The Major' was back. His career on the NCC ended when he moved on in 1941 to become Chief Officer for the LMS in Scotland. In the following years, he may have pondered just how much of a promotion this was. At York Road he was 'King', whereas in Glasgow he was just a cog in the great LMS machine. Little wonder that he ended his career in some frustration.

Speir's plans for the NCC included the construction of a direct line to eliminate the need for main line trains to run to Greenisland and reverse, the realignment of loops on the single line sections to permit fast running and mechanical tablet exchanging, and the running of faster and heavier trains to Portrush. Since the direct route over Bleach Green would involve a 3¼ mile climb at 1:76 with heavier and more sharply-timed trains, much bigger engines than any of the 4-4-0s would clearly be required.

The design of bigger engines was to lead to serious differences between Stewart and Speir. Stewart advocated compound 4-4-0s, of which the LMS then had 240 in service, while Speir was determined to have a six-coupled simple. As usual he got his way, and during a period of sick leave the unfortunate Stewart – a man in his early sixties and in poor health – received from Speir a letter which was tantamount to the sack. Railway history has more than one example of the perils of management interference in locomotive policy. The County Down Baltics have already been mentioned, and in England it was management influence that killed RM Deeley's inspired vision of a compound 4-6-0 which would have made double-heading a thing of the past. Undoubtedly though, Speir was in the right. By 1933 the gloss was wearing off the Midland Compounds.

Some years earlier, EL Diamond, in a paper to the Graduate Section of the Institute of Mechanical Engineers, exposed their weakness. This is not the place to discuss Diamond's arguments in detail, but his conclusions on the Compounds are apposite:

> At 68 mph an amount of power equal to half the work that is actually done on the train is wasted in throttling losses at this speed. In view of this, the author

unhesitatingly recommends the universal adoption of compounding as well as simple expansion of the long lap valve by means of which the port opening to steam at emission and exhaust can be substantially improved.

This conclusion was to have important consequences for design work, both at York Road and Dundalk.

The Deeley design team was divided into two camps in the late 1920s and early 1930s. JS Anderson, the Chief Draughtsman, led the advocates of short travel valves, while the rest agreed with Diamond's case for long travel with generous port openings. Anderson's view prevailed, and the Midland compounds, and most other LMS engines of the 1920s, were not nearly as efficient as they might have been.

In due course this influence spread to Ireland. When the GNRI's Dundalk team were designing their compounds, the advice they sought from Derby led them follow short travel orthodoxy, and the Great Northern Compounds had only $^{11}/_{16}$" lap for the high pressure cylinder – identical with the with the Midland Compounds. There is no doubt that if Stewart had got his way on the NCC, a 5'3" version of the Midland Compounds, with all their limitations, would have been ordered.

In his quest for a six-coupled engine, Speir had in mind the Fowler 2-6-4 tanks, introduced in 1927, which were doing fast and economical work. As he wanted an engine capable of carrying enough coal to work intensive turns of duty, the Fowler 2-6-4 tank came to Ireland in the form of a 2-6-0 tender engine. Fowler's 2-6-4 tanks had an excellent front-end design, and we may speculate on how this was achieved, given Anderson's commitment to short-travel valves. One story suggests that the valve design was done while he

was absent and preoccupied with the LMS Beyer Garratt design – an engine which was seriously handicapped by Midland ideas. There could be some truth in this since, when Anderson returned, a 2-6-2 tank emerged in 1930 which strongly resembled the 2-6-4 tank in appearance, but certainly not in the quality of front end design.

Another theory comes from OS Nock, who was present when Diamond read his paper. Nock claimed that Sir Henry Fowler was present at the meeting, and indeed had a copy of the paper sent to him. Impressed by the arguments, he ordered the drawing office to amend the front end of the 2-6-4 tanks, in the process giving the LMS its first really efficient locomotive. At any rate, the man responsible for setting out the design of the valve gear was EA Langridge, who served no less than ten Chief Mechanical Engineers, from Drummond on the LSWR to Ivatt on the LMS and the nationalised British Railways. Langridge had been told to follow the layout of the Somerset & Dorset 2-8-0s, but in fact he used the diagrams of the SECR 'River' class tanks for the LMS 2-6-4 tanks, with a 1½" steam lap.

Stewart, while still in office, worked out the broad outline of the new engines, and it is of interest that the drawings for the new engine sent from Derby bore the signature of ES Cox, one of the last great steam locomotive engineers and authors of the twentieth century. The boiler was a Derby G8S but, as explained earlier, Stanier had instructed that NCC engines should have a wider firebox to suit the more generous 5'3" gauge. Further Stanier influence could be seen in the provision of a top feed on the front ring of the boiler in front of the dome. An exhaust injector with 9mm cones was fitted on the fireman's side, and a live steam injector with 10mm cones on the driver's. There were

The first mogul, No 90, at York Road on the 9.15am to Portrush on 1 August 1933. 90 is still unnamed and has a very large 'NCC' on her tender. Despite the prestige nature of this heavy train, York Road has still managed to get a four-wheel van at the front.

William Robb

An interesting view at York Road on 14 July 1933, at the moment when the first Mogul, No 90, was delivered. On the right, in the light trousers, is a youthful Harold Houston. The excitement created by a modern outside cylinder locomotive such as this, on a railway used to conventional 4-4-0s can only be imagined. Note the absence of the coaling tower to the right of the shed (see below).

LGRP

York Road running shed, some time after the construction of the modern concrete coaling tower in 1936. The locomotives, all turned and ready to reverse down to their trains, are, from left to right, U2 class 4-4-0 No 75 *Antrim Castle*, W class 2-6-0 No 95 *The Braid*, and U2 4-4-0 Nos 86 and 85 respectively.

LGRP

9-inch piston valves with long travel of 6⅜″, and valve lubrication was by Detroit lubricator.

The balance weights on the six foot driving wheels were new to NCC practice, being built-up steel plates filled in between with the correct amount of lead. Axle boxes were cast steel with pressed-in brasses, lubricated by an auxiliary oil box on the engine frames. The leading truck was of the Bissel type, anchored to a cross-stretcher between the main frames behind the slightly inclined cylinders. A steam brake was provided on the engine with brake cylinders on the tender and steam sanding to the front of the leading driving wheels, and the front and rear of the middle drivers. The tender resembled the standard LMS Fowler pattern but was only of 2500 gallon capacity. The steam brake acted simultaneously on the engine and all the tender wheels.

The principal dimensions of the class can be found in Table 25 and a chronology in Table 26, but some comment is necessary about names and numbers. Nos 90–93 were originally to be named after famous medieval and sixteenth century figures in the history of North East Ulster (not 'Irish chieftains' as has sometimes been stated), the planned names being 90 *Earl of Ulster*, 91 *Sorley Boy*, 92 *Richard de Burgh* and 93 *John de Courcy*. However, these were not applied, though *Earl of Ulster* was later used for No 97 in 1935 (see below). Instead, No 90 was named *Duke of Abercorn* in January 1934, after the Governor of Northern Ireland, who officially opened the new Bleach Green viaduct, No 90 being used on the occasion.

Engines 91 to 95 remained unnamed until 1935–36 when they were given the names of rivers associated with NCC territory. It is worth noting that these names

Left: **No** 90 has now been named *Duke of Abercorn* and is seen at Belfast in 1936, immaculate in red livery.   *LGRP*

Opposite right: No 96, minus nameplates, at Amiens Street, Dublin in August 1936.

Henry Rea

Below: A virtually new No 92 attached to carriages in the middle road at Belfast on 2 August 1933. It received the name *The Bann* in 1936.

W Robb

Above: **No** 91, as yet unnamed, has the 'North Atlantic Express' rake in a siding at Portrush in the summer of 1935 and, from the direction the driver in looking, he is going to stable it there, ready for the next morning's 8.00am up NAE working.

Real photographs Ltd

Below: No 93, 'just out of the box' with large 'NCC' tender decals and cleaned as an engine should be, sits near the shed at Belfast in 1933. She was named *The Foyle* in 1936.

Author's collection

two round trips each day, one to Portrush and one to Larne. No 96 worked the 8.15am Belfast–Dublin and 6.40pm return. It would appear that the initiative came from the GNRI, who either officially or otherwise had heard of the moguls' performance and were anxious to try one out. The choice of a nearly-new mogul in excellent condition was made at a very high level – perhaps by Speir himself – and the NCC were clearly out to show the best that could be done by their engine.

During the interchange No 96 was driven by Con McAllister, with WR Wilson firing. Wilson's career was unusual in that it included a spell on the Portstewart steam tram, where he shared the tiny footplate with the senior driver John Burke. He finished as a senior and highly professional main line driver, a man who Mac Arnold reckoned may have been rather disappointed not to have reached the rank of Inspector after Sam Bacon's retirement.

were in the form *The Bush*, *The Bann*, etc, rather than 'River Bush', etc – the style used by the SR on the 'River' class 2-6-4Ts – or in the form used by the GNRI for the VSs in 1948 – just plain 'Foyle', 'Lagan', etc.

The 1935 locomotive interchange

In August 1935 No 96 *Silver Jubilee* (without nameplates) spent a week on the Great Northern main line, while S class 4-4-0 No 170 (then unnamed) worked on the NCC. The exchange was conducted under conditions of great secrecy, and No 170 worked

Although no official details ever appeared in print, stories filtered out that the Great Northern considered No 96 rather too small for main line express work. Log tables 19 and 20, which show a tantalisingly brief snippet of her work on the GNRI, would hardly support

Left: The moguls were mixed traffic engines and were rostered for goods as well as passenger trains. No 95 *The Braid* is pictured here at Ballymena with an up Belfast goods in 1936. Remnants of the narrow gauge can be seen to the right of the engine.

OS Nock, courtesy CP Friel

Below: No 96 *Silver Jubilee* was built in Jubilee year, 1935. She is seen at York Road in 1936. She was the engine that went to the GNRI for a week in 1936 and worked successfully on the main line. The moguls built at York Road had smaller lettering than the 1933 batch.

LGRP

this. They show No 96 making an exceptionally fast time to Howth Junction, and slightly bettering 'even time' to Drogheda. A run with Compound No 84 is included for comparison, at a time when she too was nearly new and in superb condition with the 250psi boiler. It will be noted that the compound made the faster climb of Rush bank, but the mogul accelerated away more rapidly down through Skerries.

For their part, the NCC showed no interest at all in what was a superb GNRI class, regarded by many Great Northern men as fully the equal of a compound.

The later Moguls

Nos 96–100, built in 1935–39, had names from new which were associated with the Royal Family. Rather malicious NCC gossip had it that Speir reckoned this might help him towards an eventual knighthood! No 96 *Silver Jubilee* commemorated the 25th anniversary of the accession of George V in 1910. No 97 was named *Earl of Ulster*, but this honoured the Duke of Gloucester, who visited Northern Ireland in 1935, in connection with the Jubilee, rather than the medieval Earl intended earlier. Nos 98 *King Edward VIII*, 99 *King George VI* and 100 *Queen Elizabeth* commemorated the monarchs before and after the Abdication of 1938. The only other British locomotive named after Edward VIII was GWR 4-6-0 No 6029 *King Edward VIII*, but the curious thing about No 98 was that it was named *after* he had abdicated amid some controversy.

Nos 99–104 were originally to have borne the numbers 40–45, and these numbers were stamped on to the motion parts. It was thought, however, that the numbers were more appropriate for the narrow gauge 2-4-2 tanks which were then occupying the numbers 101, 102, 104 and 111! Apart from No 103 *Thomas Somerset* (in 1942), which was named after the Chairman of the NCC, the last moguls were not named. However, under the UTA, No 101 was named *Lord Massereene* in September 1949, in honour of the last Chairman of the NCC.

Left: Even with the small tender, No 97 *Earl of Ulster* looks a tight fit on Portrush turntable 31 July 1937. Cynical NCC men suggested that Major Speirs aspired to a knighthood and that was why the later series of moguls had Royal names.

W Robb

Below: No 99 *King George VI* on 19 June 1938, just after delivery, showing that she was initially paired with a small tender.

HC Casserley, courtesy RM Casserley

Left: Now with a 3500 gallon Stanier tender, No 99 is an even tighter fit than No 97 opposite on Portrush turntable on 10 August 1938. This was a difficult turntable because of sand fouling the mechanism. Note also the Stanier chimney carried on No 99. The earlier moguls had the Fowler version (see pages 106–107).

W Robb

As they appeared, the moguls were mostly shedded at Belfast. When Coleraine received its first mogul, No 94, in 1938, its crews were not particularly happy with an engine which had an indifferent reputation in Belfast. Derry did not at first receive any mogul, as its 45-foot turntable was too short to turn one, as well as being so tight to the up main line that it needed to be interlocked with the signals. In 1936 the turntable from Cookstown Junction, a 50-foot ex-Midland Railway table, presumably regauged, was moved to Derry, and the problem solved. (This operation, incidentally, was not without incident: the new NCC steam crane was jacked into marshy ground and fell over on its side whilst trying to lift the turntable body. To the NCC's embarrassment a GNRI crane had to be brought via Cookstown to rescue it!)

The moguls were at first delivered with small Fowler tenders holding 5 tons of coal and 2500 gallons of water, similar to those on the 'Scotch' engines. However, in 1937 a much larger Stanier tender holding 7 tons of coal and 3500 gallons of water was sent from Derby to Belfast, and fitted to No 99 shortly after she was completed. This larger design was eventually fitted to the last six moguls. Firemen appreciated the shovelling plate at the same level as the fire hole (instead of slightly below it on the small tender), but complained about its draughtiness. But the necessity of such a large tender was questionable, as the moguls were never regularly called on to do the 260 mile double run to Portrush which Speir had in mind for their rostering. Besides, engine servicing at Belfast was greatly speeded after 1935 when a 'cenotaph'-type coaling plant had been built at York Road, capable of delivering two tons of coal a minute, and a 60 foot vacuum turntable installed. At the same time Coleraine, reckoned to be the principal provincial depot, also received a smaller mechanical coaling plant.

The engines were, however, called on to work some intensive rosters. The engine working the heavily loaded 9.45am to Portrush, for example, had already been to Larne and back on the 6.30am local, and the 'Portrush Flyer' engine began its roster on the 3.30am down Ballymena goods, then back on the 7.36am stopper from Cullybackey. As if this was not enough, on return from Portrush the engine set off again on another local to Whitehead at 2pm.

The Moguls: the pre-war years.

The career of the moguls can be divided into three sections, beginning with the period from 1933 to 1939. The pre-war years were an exciting time for locomotive performance, and although really high speeds were not regularly done, very fast averages on the main line were possible following the completion of the Bleach Green Viaduct and the realignment of the single line passing loops. Some of the NCC's most celebrated drivers worked the Belfast-based moguls. John Young had No 90, claiming that she was the only one of the moguls which could be safely pulled up to 5%. He later transferred to No 98, which he shared with Jimmy Marks. Con McAllister, whose name will later feature in the story of No 96's visit to the GNRI in August 1935, was the regular driver of No 95 along with Jimmy Gordon, while Bob McKenzie and Joe Kealy had No 97. Individual engine's characteristics were marked by their crews. As already noted, No 94 was considered to be a weak engine, while Nos 96 and 97 were considered particularly fast. The jumper blast pipe fitted to No 95 was reckoned to be of no great value.

The following section on the performance of the moguls refers to log tables which are reproduced on pages 118–120. Three runs in the first of the log tables illustrate something both of locomotive performance at this period, and of the wider railway context of the time. Run No 1 was recorded on the 'Portrush Flyer', whose schedule of 74 minutes left nothing in reserve for No 91 with almost a maximum load. In view of this, the climb to Ballyclare Junction was excellent, and the running over the undulating road from Cullybackey to Dunloy very good. The signal check at Ballymoney must have been extremely severe, as suggested by a time of almost thirteen minutes to pass Coleraine. The 'Flyer' returned from Portrush at 7.10pm and there was a note in the Working Timetable that, if the load exceeded ten coaches, it was to run into the Derry Central line at Macfin to clear the loop for the crossing with the 6.20pm from Belfast.

By contrast with the 'Portrush Flyer', the 'North Atlantic Express' was the NCC's lightweight showpiece train. With only six coaches, the schedule of 35 minutes to pass Ballymena and 43 minutes thence to Portrush was of no difficulty. Perhaps the most remarkable part of run No 2 was No 93's excellent minimum speed of 40mph at Mossley, and a time of under 12 minutes to pass Ballyclare Junction.

Aimed at both long-distance business commuters and holidaymakers making for the morning Stranraer ferry, the 'North Atlantic' left Portrush at 8.10am and took only 73 minutes to Belfast, including a stop at Ballymena. On Mondays to Fridays the 'North Atlantic' returned from Belfast at 5.15pm, while on Saturdays the time was advanced to 12.50pm to suit both tourists starting their holidays in Portrush and businessmen who, in those days, had to be in their offices until lunchtime on Saturday. The railwaymen knew it as "Sir Dawson Bates' train", after the Stormont MP and Minister of Home Affairs, who used it quite regularly.

'The Golfers Express' was a one-way Saturday operation in the summer months, which ran 25 minutes behind the 'North Atlantic' and offered businessmen an attractive all-in fare of 13s/6d for first-class travel to Portrush, including a meals service, plus green fees and the use of the Clubhouse at Royal Portrush Golf Club. It stopped only at Ballymoney for Ballycastle-bound golfers, who had a connecting 40 minute sprint on the narrow gauge. Portrush golfers returned on the 7.10pm up 'Portrush Flyer', whilst the Ballycastle contingent had a 9.10pm non-stop narrow gauge return special, connecting at Ballymoney at 9.55pm, with the 9.30pm ex-Portrush.

In Run No 3 (page 118), No 96 had a featherweight load on 'The Golfers Express, and had no difficulty in passing Ballyclare Junction in 13 minutes. The most remarkable feature of the run was the very high speed done through Dunadry – really fast downhill work of this sort was not typical at this period.

Runs 4 and 5 show the extremes of loading – four coaches on the 'North Atlantic' and thirteen on the 9.45am Derry express. No 96's performance certainly justified the comment made by the late 'Mac' Arnold in *NCC Saga* (Belfast 1973): "With a shorter regulator, *96* and *97* seemed almost in a little class of their own, faster though not so powerful as the Derby locomotives." Running between Glarryford and Dunloy was very fast; there was some fine speed down Ballyboyland bank and a hair-raising run through Ballymoney station and, despite a signal check at Coleraine, an excellent time of 38 minutes to Portrush – which saved exactly one minute on the schedule! Table 5 features a run with one of the heaviest trains of the day, the 9.45am to Derry. The reason why this train was so big is explained below.

The 9.45am left Belfast with the Portrush coaches next the engine, and the Derry coaches behind them. At Antrim the through coaches and vans off the 9.10am from Larne (worked round the back line from Greenisland by a Scotch engine, and arriving at Antrim six minutes before the 9.45) were standing in the up platform, marshalled in the correct order so that on attachment to the rear of the 9.45, the Larne–Derry portion was ahead of the Larne–Portrush coaches. The combined train now ran to Coleraine, where the train engine drew the Belfast–Portrush coaches forward, and a Coleraine engine backed down on to the coaches from Belfast and Larne to Derry. When the Londonderry train had departed, the Belfast engine put the Belfast–Portrush coaches on top of the Larne–Portrush coaches, which had been left standing at the Belfast end of the down platform, and then set off down the branch. And all this had to be done inside ten minutes, including multiple opening and closing of the power-operated crossing gates at Coleraine station!

No 98 clearly had the services of a banker right up to Galgorm crossing – the regulations for banking in this section have already been mentioned in an earlier chapter. With this assistance, she did well to keep this massive train rolling at over 50mph on the undulating road to Dunloy, and stopped in Ballymoney in the very creditable time of 25½ minutes.

The sixth run in this series is on the up 'North Atlantic Express', with a heavier than normal load. The work up Ballyboyland bank, and the running thence to Ballymena, was very sharp indeed. Finally comes a run in log table 7 showing the up 'North Atlantic' with two coaches more than its normal load. It is a rare example of speed not dropping below 60mph on the climb from Antrim to Kingsbog, and I would estimate that 800 dbhp was being put out for some ten

Run		1	2	3	4	5
Year		1937	?	?	?	July 1938
Train		9.20am ex Belfast *Portrush Flyer*	5.15pm ex-Belfast *North Atlantic Express*	1.15pm ex Belfast *The Golfers Express*	*North Atlantic Express*	9.45am ex Belfast
Loco		**91**	**93**	**96**	**90**	**98**
Load		320/345 tons	206/215 tons	175/185 tons	120 tons	400 tons
Miles						
00.0	**Belfast York Road**	**00.00**	**00.00**	**00.00**		
03.3	Whitehouse	06.18 45	05.35 55	06.00 52		
04.3	Whiteabbey	07.43 41	06.30 50	07.08 49		
04.6	Bleach Green	08.23 27	07.20 44	07.40 48		
05.8	Monkstown	10.46 23	09.17 42	10.00 40		
07.0	Mossley	13.05 24	10.26 40	11.15 39		
08.2	Ballyclare Junction	16.00 30	11.58 44	13.05 43		
09.2	Kingsbog Junction	17.42 41	13.50 46			
10.9	Doagh	19.37 57	15.30 66	16.05 64		
13.9	Templepatrick	22.30 70	17.55 71/72	18.50 69/77		
15.9	Dunadry	24.13 74	19.50 70			
17.3	Muckamore	25.20 71	20.57 74	21.45 75		
19.3	Antrim	26.58 72	22.21 65	23.35 60		
22.4	Cookstown Junction	29.44 60	25.20 60	26.50 58		
26.7	Kellswater	33.53 66	29.20 70	31.30 62		
31.0	**Ballymena**	38.00 54/45min	**34.38**	36.10	00.00	00.00 *banked*
33.0	Cullybackey	41.49 48/54	05.41 51	40.00 48	05.07 57	05.55
38.7	Glarryford	47.12 57/58	10.45 60	44.55 61	10.52 64	11.57 60/54
40.9	Killagan	49.27 57/60	64	47.15 63	12.55 67	14.23 54
43.6	Dunloy	52.08 63/73	15.47 60/67	51.40 *sigs*	15.23 70/75	17.23 57/72
50.9	**Ballymoney**	58.34 *sigs*	22.57 38/33	**61.00**	21.57 60/69	**25.20**
54.6	Macfin	65.34 60	27.20 53	05.10 51	25.22 70/73	05.53 54/51
57.2	**Coleraine**	71.30 56/52	33.00 47	09.45	30.04 *sigs*	**11.49**
62.5	Portstewart	**75.43**	37.36 45	13.35 52	34.21 56	
65.2	**Portrush**	**06.32**	**42.28**	**17.25**	**38.00**	

Note: Bold in all tables indicates stops.

Run		6
Year		July 1937
Train		*North Atlantic Express*
Loco		**96**
Load		220 tons
Miles		
00.00	**Portrush**	**00.00**
02.7	Portstewart	06.04 43/60
06.0	Coleraine	09.55
10.6	Macfin	14.55 61/65
14.3	Ballymoney	18.27 60
17.7	Post 50	22.23 49
21.6	Dunloy	27.03 56
24.3	Killagan	29.45 64/60
26.5	Glarryford	31.47 64
31.3	Cullybackey	35.53 67
34.2	**Ballymena**	**39.35**

Run		7
Year		July 1938
Train		*North Atlantic Express*
Loco		**91**
Load		220 tons
Miles		
34.1	**Ballymena**	**00.00**
38.4	Kellswater	05.39 70
42.8	Cookstown Junction	09.20 66/72
45.9	Antrim	11.59
47.9	Muckamore	13.46 67
49.3	Dunadry	14.59 61
51.3	Templepatrick	16.53 61
54.4	Doagh	19.50 61
55.9	Kingsbog Junction	21.23 61
57.0	Ballyclare Junction	22.22 62
58.3	Mossley	23.50 65
59.2	Monkstown	24.24 68
60.9	Whiteabbey	26.04 68
62.0	Whitehouse	27.00 74
65.2	BELFAST	**31.30** *sigs*

Above and below right: Moguls Nos 100 *Queen Elizabeth* and 101 (unnamed) at Portrush in August 1939. Both were finished in LMS red, as the NCC continued to apply red to *all* passenger engines right up to 1942. These were probably among the last photographs taken before the outbreak of war. *Author's collection*

Above left: No 103 *Thomas Somerset* was completed March 1942, still in red! She is seen at Limavady Junction on 22 April 1948, wearing post-war black livery. Note the tablet snatcher.

HC Casserley, courtesy RM Casserley

Right: Delivery of the last three moguls was delayed by the outbreak of war. No 102 (never named) was completed in April 1940. She is seen at York Road Platform 4 in UTA days on 13 July 1955.

LG Marshall

In May 1937, King George VI was crowned in London and, on 28 July, the King and Queen visited Belfast. Aside from the IRA attempt to assassinate them by exploding a bomb close by, the day brought a huge railway operation, as thousands of young people were brought to the King's Hall, Balmoral to meet the Royals. Watched by GNRI staff at Balmoral, Boy Scouts and other uniformed young people board a return special for Cullybackey, which will be routed via Lisburn and Antrim. W class 2-6-0 No 94 *The Maine* has a train composed of BCDR stock.

W Robb

minutes during this run. Remarkably, as will be shown later, work of this calibre was repeated on at least two occasions in 1965 and 1967 in the very last days of steam on the NCC. The 'North Atlantic' arrived into Belfast at a very busy time of the morning and, with three other arrivals scheduled during the ten minutes before her appearance at 9.23am, a signal check was almost inevitable. Was it, one wonders, the Cookstown passenger, which was due at 9.15am, or the 8.48 railcar from Whitehead (with bogie compo No 18 as trailer) which came in only four minutes ahead of the express from Portrush?

1939–1946: the NCC in the war years.

It was intended that all the moguls should be in traffic by 1939, but the national railway strike of 1933, and the depression years of the 1930s led to changed circumstances. It seemed, too, that Speir's estimates of traffic expansion on the main line had been rather optimistic, and that the rapid completion of the mogul order could not be justified. By 1937 York Road works were on short time, and the final batch of moguls – 99 to 104 – was not completed until 1942, rather than 1939 as at first planned. Building costs were increasing too: Nos 96 and 97, for example, cost £5164 to build, but the cost of 99 and 100 was £5913 and the last two members of the class cost £6805.

A rolling programme of boiler and firebox reconditioning began when a spare boiler was delivered along with the last mogul, No 104. This was fitted to No 93, whose boiler received a new firebox and was fitted to 92. By August 1946, when 95 had received the reconditioned boiler from 97, all eight of the original

batch of moguls had been so treated. A second new boiler – ordered as long ago as 1942 – arrived in 1946 and was given to No 94. The two spare boilers thus facilitated a series of boiler repairs and interchanges which continued until the early 1950s, when the oldest boilers became life-expired. Remarkably, only one mogul, No 103, carried the same boiler for her entire life.

In retrospect, the zenith of NCC steam was reached in 1939, but the outbreak of war caused some immediate problems. Short-time working at York Road works in the later 1930s meant that the overhaul programme was behind schedule, and the NCC was ill-equipped to handle the sudden increase in goods and government stores traffic, troop trains and numerous other reliefs. The focal point of the main line shifted from Portrush to Londonderry, with its huge defence establishments both in the city and at Lisahally.

The port of Larne became similarly important, with trains needed for the huge numbers of troops travelling on the ferries, all of which were restricted to military personnel and other permit holders. Since the Larne turntable was at the Town shed, rather than the Harbour station, paths had to be found over the one mile single-track section for engines requiring to be turned. Small wonder that during the war the NCC began to publish separate Working Timetables for passenger and goods trains! In these circumstances, punctuality took second place to the large scale movement of troops and military supplies, and the imposition of a 60mph limit on the main line underlined the message that the days of fast express running were now over.

Only nine engines had been shopped in 1937, eleven

in 1938 and ten in 1939. The situation was worsened by the slow-down of the mogul building programme during the depression years: as already noted, the last of the moguls came in 1942 rather than as intended in 1939. The final batch (Nos 101–104) performed magnificently, working far beyond the 60,000 mile deadline for works attention. Indeed No 104 ran 88,383 miles between shoppings in 1943, and No 102 an astounding 94,655 in 1945. Few other engines in Ireland would, or could, have been expected to clock up such mileages between overhauls.

Nineteen engines were overhauled in 1943, sharing space on the workshop floor with trawler and tank components which were being turned out for Harland & Wolff. The backlog built up so badly that the NCC turned to the Great Northern for help, and Dundalk overhauled Nos 15, 77, 81 and 86. It has already been noted that, in a generous gesture, the Great Northern returned No 81 in the full NCC lined maroon livery – a contrast with the plain black (if painted at all) of engines emerging from York Road during the war. Even with the GNRI's help, the NCC had reached breaking point by the end of the war, and Harland & Wolff's shipyard finally stepped in to overhaul moguls Nos 90, 91, 92, 93, 95, 96, 100, 102 and 104 (not, of course, in numerical order!) as well as 'Scotch' engine No 72. It will be noted later that Nos 91, 94, 98 and 103 were overhauled by Harlands in 1949, meaning that eventually all but three of the class – Nos 97, 99 and 101 – visited the shipyard for overhaul.

As a firm of shipbuilders, whose only railway construction work was a few diesel locomotives, Harland & Wolff were not fully conversant with steam locomotive overhauls. In particular, setting valves is a tricky task and the attempts of Harland and Wolff were not likely to please NCC drivers, especially if their pet engine was involved.

Some idea of the work the NCC locomotive department was expected to handle, can be gleaned from a study of the locomotive rosters for the year 1943. NCC engines were arranged in power classes, alphabetically graded beginning with A, the engines with the smallest tractive efforts. The seven-foot compound No 55 *Parkmount* was the only engine left in this category after the scrapping of F class No 23 and C class compound No 56. Power Class B comprised the 'Whippets' of both 7'0" and 8'2" wheel bases, and the surviving K class, No 31. In Class C were the U1 and U2 class 4-4-0s, while Class D compromised the V class 0-6-0s. The moguls were in the most powerful class, E.

The surviving class A engine, No 55, spend most of the war in Belfast passenger yard on Turn 9, as well as sharing goods and passenger to Ballyclare and Cookstown with the Class D 0-6-0s. The Donegall Quay shunts (Turns 12 and 13) were covered by 'The Donkey' – the faithful 0-4-0ST No 16. The Larne line was covered by the Class C engines, which worked turns of up to 23¾ hours and were washed out at Cookstown Junction and Coleraine sheds, as well as Larne Town. Even the most venerable Class B engines were not spared – Turn 38 was a 19¼ hour day based on Derry, while a Cookstown Junction engine had a 14¾ hour day and in Turn 31 a Class B engine shunted for 21¾ hours with three different crews.

The four biggest turns for the moguls were numbered 1 to 4. Turn 1, Coleraine-based with 26 hours in traffic, led to Turn 2, a Belfast job of 375 traffic miles and 22 hours work. This led on to Turn 3 in Coleraine with 316 miles in 19 hours, and then Turn 4 with 261 miles in 16½ hours. The moguls worked these turns in rotation and were serviced and washed out at Coleraine. Not

Above right: The last wartime mogul was No 104, completed in October 1942. Here, she works the up Derry express through Limavady Junction in 1955. Note the outside steam-pipes and the beautiful array of signals, including a gantry at the Londonderry end. The Limavady branch trailed to the left behind the platform.

Author

Run		8	9
Date		11 July 1940	?
Train		5.55am ex Belfast	?
Loco		**102**	**101**
Load		330 tons	310 tons
Miles			
00.0	**Belfast York Rd**		**00.00**
03.3	Whitehouse		06.30 45
04.3	Whiteabbey		08.02 38
04.6	Bleach Green		
05.8	Monkstown		11.20 37
07.0	Mossley		13.30 28
08.2	Ballyclare Junction		16.03 30
09.2	Kingsbog Junction		17.53
10.9	Doagh		20.02 56/67
13.9	Templepatrick		23.13 63
15.9	Dunadry		25.07 70
17.3	Muckamore		26.24 63
19.3	**Antrim**		**28.45**
31.0	**Ballymena**	**00.00**	
33.0	**Cullybackey**	**07.30**	
38.7	**Glarryford**	**08.49** max 49	
40.9	**Killagan**	**04.58** max 37	
43.6	**Dunloy**	**05.42** max 40	
	Post 48	04.02 64	
50.9	**Ballymoney**	**10.27**	
54.6	Macfin	06.11 57	
57.2	**Coleraine**	**12.32**	

Run		10
Date		7 Nov.1942
Train		?
Loco		**103**
Load		360 tons
Miles	(from Portrush)	
34.1	**Ballymena**	**00.00**
38.4	Kellswater	07.15 55/50
42.8	Cookstown Junction	12.15 54/60
45.9	**Antrim**	**15.50**
47.9	Muckamore	05.55 38
49.3	Dunadry	08.15 34
51.3	Templepatrick	12.01 36
	Post 15	13.55 36
54.4	Doagh	16.55
55.9	Kingsbog Junction	19.10 39
57.0	Ballyclare Junction	20.52
58.3	Mossley	21.56 57
59.2	Monkstown	23.04
60.9	Whiteabbey	sigs
62.0	Whitehouse	26.30
65.2	**Belfast York Rd**	**34.15**

all the work done in these turns was glamorous by any means. The moguls on Turns 2 and 3 could be found in the unusual environment of Larne Harbour, working through to Coleraine, and on goods trains to and from Ballymena and on to Coleraine. One mogul brought a goods from Coleraine to Belfast, then ran light engine to Ballyclare Junction to work a further goods to Larne.

Logs recorded during the war are scarce, partly because of the difficulties of travelling, and partly because of the sensitivity of many of the locations which the NCC served; photography was illegal, and furtive note-taking would have been viewed with great suspicion! Run 8 shows a brand-new mogul on the one train of the day that served all stations between Ballymena and Ballymoney, while Run 9 was particularly good from Whiteabbey to Ballyclare Junction. Despite the 60mph wartime limit, the driver allowed No 101 to run away to a brief 70mph. Run 10 shows a really big load being competently handled up the hill out of Antrim. A sustained speed in the mid-thirties with such a heavy train would indicate an output of around 800 dbhp, or 36hp per square foot of grate. Outputs of this order certainly show that Speir was justified in his insistence on a six-coupled engine rather than a 4-4-0 compound in 1932!

It might also be noted that when King George VI and Queen Elizabeth visited Northern Ireland in 1945, the appropriate engine, No 99 *King George VI*, was awaiting shopping, so No 101 was given the *King George VI* nameplates for the occasion and the train was worked by senior driver Con McAllister. No 101 was also repainted into pre-war maroon livery, and lined out in two days, for the Royal journey from Lisburn to Lisahally, the last NCC engine to emerge from the works in this livery. This brought to an end a beautiful and distinctive livery, which had rather surprisingly been applied to the three wartime moguls (102–104), as well as to 'Scotch' engine No 81 on the occasion of her wartime overhaul at Dundalk works by the Great Northern. However, most other NCC locomotives had received plain black in the later war years.

The Postwar period: 1945–1949

The NCC emerged from the war with its York Road headquarters badly damaged, much of its rolling stock destroyed and its priceless drawings and records burned. Remarkably, not a single engine was damaged, and with the aid of a further contract with Harland and Wolff in 1949 to repair Nos 91, 94, 98 and 103, the moguls were gradually restored to sound mechanical order. The UTA was now in control of both the former

Left: A 1948 view of No 100 *Queen Elizabeth* as fitted for oil burning in October 1947. She is in the post-war black livery.
Real Photographs Ltd

Below: No 97 *Earl of Ulster* crosses the River Bann using the Curragh bridge on the Derry Central in 1947, as it heads for Kilrea.
Kenneth Benington

Right: A busy day at Ballymoney on 18 April 1948, as No 104 with a down train, crosses No 97 on an up express. In the background, narrow gauge 2-4-2T No 43 is about to reverse down to her train on the other side of the island platform.
HC Casserley, courtesy RM Casserley

Run		**11**	**12**
Date		25 May 1953	1 Aug 1955
Train		5.30pm ex L'derry	5.30pm ex L'derry
Loco		**91**	**104**
Load		260 tons	260 tons
Miles			
00.0	**Londonderry**	**00.00**	**00.00**
04.9	Culmore	07.50 60	08.23 56
07.6	Eglinton	10.29 65/30*	11.18 61
12.9	Lisahally	15.36	16.08 63/66
13.6	Ballykelly	17.24 50	17.05 66
18.6	**Limavady Junct**	**20.10**	**19.16**
20.4	**Bellarena**	**07.08**	**07.51**
23.4	Magilligan	05.01	05.11
26.4	Downhill	07.48 60/65	08.15 56
27.8	**Castlerock**	**10.03**	**10.19**
32.3	Post 63	05.58 61	06.17 59
33.6	**Coleraine**	**08.43**	**08.19**

Run		**13**	**14**
Date		5 May 1952	22 April 1953
Train		5.25pm ex Belfast	5.25pm ex Belfast
Loco		**102**	**100**
Load		260 tons	300 tons
Miles			
00.0	**Belfast**	**00.00**	**00.00**
03.3	Whitehouse	05.58 53	07.28 41*
04.3	Whiteabbey	07.17 50	09.00 38
05.8	Monkstown	09.58 36	12.20 26
07.0	Mossley	11.41 33/30	14.21 27
08.2	Ballyclare Junct	14.01 30	16.56 30
09.2	Kingsbog Junct	15.36 29	18.44
10.9	Doagh		20.15 62
12.5	Post 15	18.50 70	21.46 78
13.9	Templepatrick	20.00 74	22.52 79
15.9	Dunadry	21.40 76	24.23 82/85
17.3	Muckamore	22.48 69	25.26 75
18.5	Post 21	23.55	26.24 60
19.3	**Antrim**	**25.08**	**27.39**

LMS(NCC) and the BCDR, so the chance was taken to run each of the moguls in on the Bangor line, certainly the biggest tender engines that this busy commuter line had ever seen. No 103 had the most flying visit of them all, working only a few trains before being rushed back to York Road to work the inaugural 'Belfast Express' from Derry on 5 July 1949.

Right at the end of the NCC regime, Nos 100 and 101 were briefly modified as oil-burners. Their regular drivers, Joe Shiels and Jack Redfern, were agreeable enough to the experiment. No 103 was considered for conversion, but no-one would have dared to suggest taking the newest engine away from its regular driver Harry Molloy!

Not long before his death, former driver Sammy Sloane recorded his memories of firing the oil burners – at which time he would have been perhaps the last man alive to have worked regularly on them. As fitted for oil burning, Nos 100 and 101 were not universally popular, and Sammy would often get an extra turn, in place of regular firemen who preferred the shovel.

The oil was heated by copper spirals in the tenders, and it needed 100psi pressure to keep them hot enough to make the oil flow. There was a bracket for the projection plate on which the three jets were mounted. Each could be adjusted for oil flow and steam pressure, and this required careful management to avoid jets of burning oil flowing through the grate; there was at least one occasion when the sleepers at Cullybackey station were set on fire! There was firebrick under the brick arch, and when the oil jets were shut down, the brick surface

was covered by what looked like a red hot balloon of oil. Grover, the patentee of the device, advised that the jets be cleaned nightly before they cooled down. As there was no blower on the oil burners, blow backs were not uncommon, and enginemen could finish their day as black as sweeps!

Sammy fired to Joe Shiels on a ten-coach test run with No 100, in the presence of no less than RA Riddles himself. Riddles reckoned the jets were not burning right, and persisted in lifting the flap of the door (the lower part of which was welded up) to inspect them. Not a bit overawed by the great man, Shiels called out; "Keep that flap shut would ye! Ye're lettin' the coul' air in!" During the course of the run one of the oil pipes broke, and at Portrush Riddles drily remarked that it wasn't often he'd seen time being kept on an express train with less than 100psi of steam. The NCC was not really serious about oil burning, and oil of an inferior nature was brought in by road tanker. The consensus among enginemen was that the engines were no more powerful than coal-burners and that, since the oil-burners did only one out-and-back trip to Portrush daily, there was little economy either.

Recovery in services was almost as swift as its descent had been six years previously, and in little more than a year new timetables restored much of the interest in NCC performance. The emphasis was still on serving Derry, and Portrush never regained its predominant position. The 8.25am and 5.25pm expresses were booked for 2¼ hours, including stops at Ballymena, Ballymoney, Coleraine, Castlerock and Limavady Junction, and the corresponding up trains

Left: The first livery applied to the moguls by the UTA after 1949 was black with large 'U T' letters on the tender. No 102 is seen in this style with a down train near Ballymoney. The first five carriages on this train are the famous 1934–35 'North Atlantic Express' rake.

Kenneth Benington

Right: Another view of No 102, about 1950, this time with an up express between the two tunnels at Downhill, west of Castlerock, with the Atlantic in the background. Only Nos 102 and 103 appear to have received this livery, the UTA roundel with red hand emblem being introduced almost immediately.

Kenneth Benington

Left: Mogul No 103 *Thomas Somerset*, also with the 'U T' lettering, climbing away from Ballymoney with a heavily loaded up express in the summer of 1950.

Kenneth Benington

In May 1951 the UTA commemorated the Festival of Britain by introducing 'The Festival' express between Belfast and Londonderry and introducing new steel-clad corridor stock. No 94 *The Maine* approaches Ballymoney with a lightly loaded up 'The Festival'
Kenneth Benington

left Derry at 8.30am and 5.30pm The pre-war 'North Atlantic' set with one additional coach made up one set, while the best of the remaining vehicles made up the other. Moguls provided the motive power for these trains, and for the first time one engine worked through from Belfast to Londonderry. But pressure grew for through coaches to Portrush, so the pre-war practice was restored, and loads again increased to eight or nine bogies. There was an advantage in economy for the railway too – by reverting to a Belfast–Portrush trip for the Belfast crews, mileage money (for which 140 miles was the qualifying distance) was avoided.

Under the UTA: 1949–1958

The early UTA period brought few changes at first, other than experiments with liveries. At least two moguls, Nos 102 and 103 *Thomas Somerset*, received large 'U T' tender decals on black livery and No 90 *Duke of Abercorn* was given a Brunswick Green livery (as depicted on page 16 of Derek Young's *The Ulster Transport Authority in colour*), before management came down firmly against anything but black.

In 1951, the Government staged a 'Festival of Britain' at London and the UTA joined in the celebrations by introducing a new named train between Belfast and

Londonderry called 'The Festival'. This train was almost invariably hauled by moguls and began operation on 3 May 1951, with a brand new set of carriages, hauled on the first day by No 99 *King George VI*. The up working was the 8.30am ex-Londonderry, the balancing working from York Road being the 5.25pm. There was a stop at Antrim and a three coach Portrush portion was detached at Portrush, all very tight for a two hour, 15 minute, start to stop schedule.

Log tables 11 and 12 show two good examples of the best work being done at this period on the 5.30pm Derry–Belfast express. Each was checked at Bellarena to cross the late-running 3.40pm from Belfast; had it not been for these checks, both Driver Mills of Coleraine and Driver Hinds of Derry would have kept that difficult 50mph booking from Limavady Junction to Coleraine over track which is pretty well level throughout. When 'Mac' Arnold wrote *NCC Saga*, the definitive study of NCC operation in the postwar years, he entitled the chapter covering 1952–1954 'The Great Mogul' – an appropriate title indeed! Some of that greatness can be seen in two quite exceptional runs by Coleraine drivers Millar and Mills, detailed in log tables 13 and 14. With 260 tons, No 102 did well to fall no lower than 30mph at Ballyclare Junction, and there may have been a distant

Right: Mogul No 95 *The Braid*, wearing the new UTA roundel, steadily climbs Ballyboyland bank with a heavily loaded up express about 1951. Five of the first seven carriages are from the new 'Festival' stock.
Kenneth Benington

Below: In Coronation Year, the new Queen and the Duke of Edinburgh visited Northern Ireland. A special joint UTA/GNRB Royal Train was prepared, using six GNRB coaches and four UTA. All were painted into the GNRB Oxford blue and cream livery. As No 100 *Queen Elizabeth* was not in good enough condition, No 102 did the honours and, in this view, approaching Ballymoney on 3 July, she wears the UTA roundel.

Kenneth Benington

Opposite bottom: No 91 *The Bush* passing Lurgan with an up train about 1961. In early 1959, she was the first mogul to arrive on the GNR. The covered footbridge dominates the scene and the numerous bike sheds, a feature of Lurgan at the time, can also be seen. No 91 was one of the first four of the class and was one of the last to remain in service.

Author

signal sighted at Kingsbog. The run was done in two minutes under the 27 allowed. No 100's performance, with Sammy Mills of Coleraine, striving hard to regain time after that temporary speed restriction at Whitehouse, resulted in the highest known speed with a mogul, and a particularly fast approach to Antrim.

Nor was management particularly sympathetic in these days. Intended to produce a co-ordinated road and rail transport network for Northern Ireland, the bus-minded Ulster Transport Authority lost little time in closing most of the BCDR and pruning the NCC quite viciously too. In addition, from the mid 1950s steam was phased out in favour of diesel multiple units, which were less comfortable and less mechanically reliable than the superb steam locomotives they replaced.

Not only did the moguls face an unsympathetic

owner but, from 1946 onwards, they faced competition from younger rivals, the WT class 2-6-4 tanks. Indeed from 1950 onwards, at least one mogul would be greased up and put into winter storage, often at Whitehead or Cookstown Junction sheds.

Shopping and boiler work continued during the 1950s. In 1952, No 104 went into the shops, gaining outside steam pipes, new cylinders and the boiler off No 102, along with a reconditioned firebox. She was an indifferent steamer however and eventually Billy Hanley had the orifice of her blastpipe altered from 4¾" to 4⅝".

A further Royal train was run in July 1953, from Lisburn to Lisahally, for the visit of the newly-crowned Queen Elizabeth II. This was a ten-coach train with four UTA vehicles and six GNRB. No 100 *Queen Elizabeth* would have been appropriate motive power but was not in good enough condition, so on 3 July 1953 driver Joe Shiels and fireman Percy Mitchell had No 102, immaculately turned out in UTA black with the attractive red and straw lining. On this occasion, nameplates were not transferred (see page 113). The composition of the train was as follows (UTA vehicles in bold): 231-403-227-**3**-50-**8**-87-225-**90**-232. Of these, 231 and 232 were brake firsts, 227 and 225 corridor firsts, 403, **87** and **90** dining cars, and **3**, 50 and **8** saloons. The entire train was painted in the GNRB Oxford blue railcar livery, and the UTA vehicles stayed in these colours for several years.

The year 1954, though, marked the withdrawal of the first engine of the class. No 90 by this time had 96's original boiler, and in this form was withdrawn, being scrapped two years later. Nos 91 and 93 were reboilered at this period, 91 gaining 104's reconditioned boiler and No 93 the boiler off 99, along with a reconditioned firebox. Between then and 1961 only one other mogul, No 100, had a boiler change.

The culling of the moguls began in earnest in 1956, with Nos 96, 101, 102 and 103 all steaming for the last time. No 101 caused some surprise, since she had last received works attention only seven months before she was suddenly withdrawn. Her firebox still had plenty of life, though her boiler – second hand from No 95 in 1947 – was over the regulation 20 years. One story had it that the tubeplate was condemned, another that the engine had been withdrawn in error. The end result, sadly, was not in doubt. Nos 96 and 103 were retained

in storage for some time, but Nos 90, 101 and 102 were cut up in 1956.

They were followed in 1957 by No 92 and in 1959 by Nos 100 and 103. No 96 was cut up in 1961. The remainder of the class, as will be seen in the next section, were to spend their final years on the Great Northern after 1958 – the year when the UTA took over the Northern Ireland lines of the former GNRB and dieselised the principal services on the NCC main line. The last appearance of a mogul on the NCC was during the June 1964 IRRS/RCTS/SLS joint 'Farewell to Steam' Irish rail tour when No 97 *Earl of Ulster*, at that time based on the Great Northern, worked over the Antrim branch from Great Victoria Street and through to Portrush and on to Londonderry.

The Moguls on the Great Northern

The Great Northern Railway Board had, since 1953, been the joint responsibility of the governments of Northern Ireland and the Republic of Ireland. The Northern Ireland government's refusal to assist with the GNRB's growing financial crisis led to the closure of much of its route mileage in Counties Armagh, Fermanagh and Tyrone, and in 1958 its independent existence ended with a split between the UTA and CIE. There was still healthy traffic on both the main line and the Great Northern 'Derry Road' and, much as the UTA would have liked to close even more, they faced a short-term locomotive shortage following the 50:50 division of the locomotive fleet between the

Northern and Southern state transport undertakings.

The solution was to transfer both moguls and 'Jeeps' to the Great Northern. Over the next few years Nos 91, 93, 94, 95, 97, 98, 99 and 104 become familiar engines at Adelaide and Portadown sheds. No 95 had received her last overhaul in 1959, followed by No 93 in 1960, Nos 98, 99 and 104 in 1961, No 94 in 1962 and Nos 91 and 97 in 1963. Sadly, none of these engines ran high mileages after their last shopping. No 95 was stopped by 1961, Nos 94 and 98 by 1963, and Nos 95 and 98 were broken up in 1964.

The others just about made it into the last full year of regular steam – 1964. Nos 99 and 104 had in fact been stopped early in the year, but were reinstated for the final steam summer timetable before the 1965 closures. Only No 97 made it through to the closure of the 'Derry Road' at the beginning of 1965. Five of the last six moguls, Nos 91, 93, 94, 99 and 104, were sold for scrap in May 1965, but No 97 was retained for another seven months, and was rumoured to have been considered for preservation before scrapping in December 1965.

There are a number of reasons for this rather depressing end to the mogul era. Certainly they arrived on the Great Northern at just about the worst psychological time. Enginemen are by nature conservative and suspicious of anything new – witness the hostile reaction of Great Northern men when the Compounds arrived in 1932, and the NCC men when the 'Jeeps' came in 1946.

But in the case of the moguls there were other factors. For one, the engines were not in first class condition – if they had been, they might have been given a fair trial. Worse still, they were seen as cast-offs from an organisation which had broken up 'their' railway and seemed to have no interest in the Great Northern and its traditions. For men facing an uncertain employment future, rough-riding engines with a steam brake rather than the Great Northern vacuum brake were about the last thing they would welcome. Their continuous use on goods train over the Great Northern 'Derry Road' battered their already weakened frames. (It will

Moguls Nos 98 and 104 were fitted with outside steam pipes just after the war. No 98 is seen at Portadown Junction with an ex-GNR carriage and a mixture of cattle wagons and horse boxes on a special conveying bulls to the RDS in Dublin. *Author*

No 99 at Dundalk, home of the GNRI, starting the climb of the Wellington bank. A typical GNRI bracket signal can be seen behind the water tank Another almost forgotten occupation is the permanent way man in hat, with hammer and spanner, standing on the right.

Author

be remembered that Glover had once said that no engine should do more than six months at a time on this difficult route). And on the main line, a run down mogul was no substitute for a Great Northern VS class 4-4-0 on non-stop Belfast–Dublin specials.

The picture was not all gloom, however. The younger Adelaide and Portadown drivers conceded that, on its day, a mogul 'could run', and was an economical steamer. The well-designed front end ensured that a mogul could get along effortlessly on 15–20% cut off and had Dundalk produced a small mogul at the time they brought out the first of the UG class 0-6-0s in 1937, it would surely have been a success. Laurence Liddell's assertion that "No single Great Northern engine could

Run	**15**	**16**
Date	24 Feb 1962	6 March 1964
Train	9.25am special	9.25am special
Loco	**93**	**97**
Load	Ten bogies 310/320tons	Seven bogies 220/240 tons
Belfast Gt Victoria St	**00.00**	**00.00**
Dunmurry	09.59 34	08.41 40
Lambeg	13.05 41	11.16 48
Lisburn	15.20 37	13.06 47/58
Damhead	21.32 55	19.04 34 *signals*
Moira	24.29 59	22.02 29 *bridge reconstr*
Post 96	26.00 46	25.25 39
Lurgan	29.30 61	29.03 63
Boilie	31.22 67/68	30.55 66
Seagoe	33.15 60	32.53 49
Portadown [res 25]	34.33 33	34.19 27
Tanderagee	42.42 53	41.42 59
Scarva [res 50]	45.25 52/55	44.25 53
Poyntzpass [res 40]	50.15 15 *signals*	47.26 44
Post 74	56.20 47	50.56 62
Goraghwood	61.20 *signals: near stop*	53.19 44
Bessbrook	67.44 28/24	56.50 40
Bridge 180	77.15 33	25 min
Meigh	63/61	60 max
Adavoyle	80.35	68.49 55/53
Post 57	85.48 70	7423 66
Dundalk	**89.38**	78.41 33
Post 52		82.47 36
Castelbellingham		88.05 56/50
Dromin Jcn		92.08 52
Dunleer		94.19 54
Kellystown		110.09 41/56
Drogheda [res 15]		107.45 19
Laytown		114.00 62
Mosney		115.15 58
Gormanston		117.06 58
Skerries		123.27 60
Post 16		125.33 52
Rush & Lusk		127.49 54
Donabate		130.23 57
Malahide		133.21 43
Portmarnock		135.54 56
Howth Junction		137.58 52/46 *signals*
Killester		140.55 51
Post 1		142.57 17 *sigs*
DUBLIN AMIENS St		**146.55** *Arrive 3 min early*

Run	**17**
Date	July 1963
Train	?
Loco	**93**
Load	Two bogies (60/65 tons)
Lurgan	**00.00**
Kilmore	04.10 54
Post 96	05.20 53
Post 97	06.15 66
Moira	07.40 75
Damhead	09.07 78
Post 101	09.55 75
Post 102	10.40 79/80
Knockmore	12.00 75
Post 104	12.30 66
Lisburn	**13.50**

Run	**18**
Date	1 August 1964
Train	5.05pm ex Omagh
Loco	**97**
Load	Six bogies
Omagh	**00.00**
Post 41	02.41 31/36
Post 40	04.24 33
Post 37	08.34 46
Post 36	09.54 46
Beragh	12.17 23 *hand exchange*
Sixmilecross	14.59 46/47
Post 30	18.41 36
Carrickmore	20.11 43
Post 27	23.35 29 min
Post 25	27.06 40
Pomeroy	28.41 28 *hand exchange*
Post 21	33.11 51
Post 18	36.51 39
Donaghmore	37.17 42
Post 17	38.21 40/47
Dungannon	**41.41** Allowed 47
Post 12	05.25 53
Trew & Moy	06.41 20 *hand exchange*
Vernersbridge	09.34 39
Post 8	11.14 50
Annaghmore	13.17 28 *hand exchange*
Post 4	17.19 48/51
Portadown Junct	21.56
PORTADOWN	**23.34** *Allowed 25*

Run	19	20
Date	1935	1932
Train	?	?
Loco	96	84
Load	290/310 tons	290/310 tons
Dublin Amiens St	**00.00**	**00.00**
Clontarf	03.35 30	04.15 24
Howth Junction	07.40 60	08.12 58
Malahide	11.25 74	11.56 68
Donabate	13.25 71	13.58 71
Rush & Lusk	15.26 70	16.03 72
Post 16	17.38 58/76	18.06 61
Skerries	19.18 75	19.52 64
Balbriggan	22.35 74	22.54 75
Laytown	26.57 75	27.10 75
Drogheda	**31.45**	31.30 pass

have efficiently worked both the 'Enterprise' and the 'Porter Train' as a mogul could surely have done" is a thought-provoking comment about their unfulfilled potential on the Great Northern.

On the Great Northern the moguls were rated as 'C class', along with the SG and SG2 0-6-0s which were allowed 47 wagons (450-500 tons gross) from Dundalk to Goraghwood. Here it is apposite to consider the relative difficulties of the climbs from Dundalk to Adavoyle and from Belfast York Road to Kingsbog Junction. On the Great Northern, the line rises 325 feet in seven miles, with long stretches of 1:100, while the NCC has a gentler rise of 312 feet in 9½ miles, but a steeper ruling gradient of 1:75 for three miles.

In their declining years, the moguls gave a good account of themselves on trains like the 12.45pm Dundalk–Portadown goods, able to keep a train moving on the level around Tandragee with the cut off set at around 15%. This gives substance to RN Clements' opinion that an NCC mogul could be more expansively worked than any other engine in Ireland. An abiding personal memory of a mogul on Great Northern goods work is of the 9.50pm Derry goods. The late Drew Donaldson and I went out to milepost 26, and I can still remember the approaching sound of No 97 blasting away from Pomeroy on that clear and frosty night, her exhaust crackling like a machine gun as she pounded past us on her way up Carrickmore bank.

On passenger work, the moguls were mostly to be found on trains that would scarcely have taxed them in their heyday: lightly loaded local trains between Belfast and Portadown and on the Derry Road, interspersed with occasional appearances on the Dublin semi-fast trains. In the last summer of steam on the Great

Above: No 94 *The Maine* leaving Dungannon in 1963. The GNRI signals on the left are very different in pattern to the somersaults of the NCC. Lack of cleaning shows on the engine's shabby appearance.
A Donaldson

Left: Goraghwood was a favourite photographic spot and a grimy No 97 *Earl of Ulster* is pictured starting the climb to milepost 65¼ with the 12.30pm to Dublin on 14 July 1964. This climb was longer than the Greenisland loop line, though not as steep, and could tax a shy-steaming engine.
Author

Northern, No 97 was based at Adelaide and appeared on both Dublin excursions and such trains as the 12.30pm Belfast–Dublin as far as Dundalk, along with the more run-down Nos 99 and 104 on such trains as the Omagh relief and 10.15am from Derry.

The major debating point in relation to passenger trains was whether a mogul could comfortably work non-stop to Dublin with heavy special trains in connection with Rugby Internationals at Lansdowne Road. Log table 15 shows the result of one unsuccessful attempt with No 93 and a ten-bogie train. Billy Hanley went with driver Arthur Boreland of Adelaide – an intelligent and economical engineman who, it will be seen, was also involved in a more successful attempt two years later. No 93 was, unfortunately, badly blocked behind a preceding special with Great Northern S2 class 4-4-0 No 63 *Slievenamon* (ex-192). There were two severe checks at Poyntzpass and Goraghwood and, when the train was again brought into Dundalk under caution, the decision was made to take water. The rest of the run was very badly blocked all the way to Dublin.

Log table 16 shows how, on 6 March 1964, with a lighter load and NCC Inspector Frank Dunlop, Boreland managed the non-stop run without difficulty, and just about within the 2 hours 25 minute allowance. This ended attempts at non-stop runs, though in 1987 the RPSI celebrated the 40th anniversary of the 'Enterprise' express by staging a non-stop run from Belfast Central to Dublin with preserved GNRI compound No 85 *Merlin*.

Log table 17 shows something rather different, with No 93 on a featherweight two-bogie train. This appears to have been a two-bogie scratch relief run to convey passengers to the Liverpool steamer, due to the late running of the down evening Dublin service. The mercurial Barney Burke of Portadown produced an electrifying run, with one of the last displays of speed in mogul history.

Other turns of speed were not uncommon. On 11 April 1964, for example, Driver Victor Cust of Portadown had No 97 and a four coach train, running from Belfast to Lisburn in 10'32" with a 60mph maximum at Lambeg, and two weeks later No 91, on the 2.45pm from Dublin with six bogies, ran from Goraghwood to Portadown in 18'52", with 69mph maximum after Tandragee. Log table 18 shows a further bright run with Driver Alec Murdy of Adelaide on the 5pm Omagh–Belfast summer relief train, with a good effort over Carrickmore bank.

Table 25: Dimensions of the W class 2-6-0s, built 1933–42

Type	2-6-0
Cylinders	19"x26"
Piston valves	9 inches
Coupled wheels	6' 0"
Bogie wheels	3' 0"
Wheel base	9' 0" + 8' 0" + 8' 6"
Boiler length	10' 10½"
diameter	4' 6⅞"
tubes	121 x 1¾"
Superheater tubes	21 x 5⅛"
Heating surface	1347 sq ft
Firebox	7' 6"
Grate area	25 sq ft
Boiler pressure	200 psi
Tractive effort	22,160 lbs
Weight	62 tons, 10 cwt
Tender weight	32 tons, 19 cwt
Water capacity	2500 gallons
Coal capacity	5 tons

Notes: Nos 99–104 had 3500 gallon tenders, with 7 tons of coal, weighing 47 tons 16 cwt.

Nos 99–104 had 138 small tubes with HS 1437 sq ft.

Table 26: Chronology of the W class 2-6-0s, Nos 90–93 built at Derby and Nos 94–104 at York Road

No	Name	Deliv	Cost	Boiler changes	Last overhaul	Last used	Final mileage	Scrapped
90	Duke of Abercorn (1/1934)	7/1933	£6617	1946? (ex-96)	1949	7/1953		6/1956
91	The Bush (11/1935)	7/1933	£6617	1946 (ex-92?) 1954 (ex-104)	1/1963	9/1964	1,407,695	5/1965
92	The Bann (1/1936)	7/1933	£6617	1943? (ex-93)	?	7/1956		10/1957
93	The Foyle (2/1936)	8/1933	£6617	10/1942 (new) 1954 (ex-99)	10/1960	12/1963	1,377,826	5/1965
94	The Maine (2/1936)	6/1934	£5707	1946 (new)	10/1959	3/1964		5/1965
95	The Braid (3/1936)	10/1934	£5707	8/1946 (ex-97)	1959	11/1960	1,159,274	8/1964
96	Silver Jubilee	5/1935	£5164	194? (ex-91?)	9/1953	5/1955		12/1961
97	Earl of Ulster	7/1935	£5164	1946	10/1963	12/1964		12/1965
98	King Edward VIII	2/1937	£5950	6/1949 (ex-100)	1961	11/1963	1,070,136	8/1964
99	King George VI	5/1938	£5912	1951 (ex-101) 1961 (ex-93)	1961	10/1964	1,168,272	5/1965
100	Queen Elizabeth	1/1939	£5912	1947 (ex-94) 1955 (ex-102)	12/1955	9/1957		12/1959
101	Lord Massereene (9/1949)	6/1939	£5977	1947 (ex-95)	12/1953	3/1956		6/1956
102	–	4/1940	£6028	8/1950 (ex-98)	4/1953	1/1956		6/1956
103	Thomas Somerset	3/1942	£6806	never	2/1954	8/1956	696,750	12/1959
104	–	10/1942	£6806	10/1952 (ex-102)	6/1961	4/1964	830,133	5/1965

Notes: Nos 90–93 were originally to have been named as follows:

No 90 *Earl of Ulster*   No 91 *Sorley Boy*   No 92 *Richard de Brugh*   No 93 *John de Courcy*

No 101 were originally to have been named *Lord Massereene and Ferrard*.

Nos 99–104 were originally intended to have been Nos 40–45.

Nos 99–104 were given Stanier 3500 gallon tenders from new, though No 99 ran briefly with a Fowler tender.

Stanier tenders from Nos 101 and 102 were transferred to Nos 91 and 97, respectively, in 1956.

The Stanier tender from No 103 was transferred to No 95 about 1958 and probably went to No 94 about 1961.

The Stanier tender from No 100 was transferred to No 96 about 1959 and then to No 98 in late 1961.

No 103 was named after the Chairman of the NCC Board of Directors.

Chapter 9
The Jeeps

It was significant that the last class ordered for the NCC should, at eighteen engines, share with the U2 class the distinction of being the most numerous. In the same way that Speir had campaigned for the introduction of the moguls, his successor, Major Frank Pope, seems to have been the driving force behind the 'Jeeps'. A man whose railway career had been spent on the LMS, and who had experience of railways and docks in India in the 1930s and 1940s, would have regarded the NCC as a small system which could be adequately worked by tank engines. He reckoned that eighteen new tank engines, plus existing machines rebuilt as tank locomotives, could run all the traffic on the NCC, and York Road station was to have been redesigned to incorporate platform run round roads.

The design was to be based on the LMS 4P design, but modified so as to be standardised with the moguls. The idea of a Derby-designed 2-6-4T can be traced back to as early as February 1914, shortly after the Midland Railway had taken over the London, Tilbury and Southend Railway. The LTSR had recently introduced 4-6-4 Baltic tanks which the Midland considered too heavy. The Derby proposal was for a 2-6-4T with 6'3" driving wheels and cylinders and valve gear based on the S&DJR 2-8-0 design. Elements of this proposal actually appeared in the 1917 'River' class 2-6-4Ts

of the SE&CR because draughtsman James Clayton moved from Derby to Ashford in March 1914 to serve under Maunsell and took a set of drawings with him.

The war then intervened and it was some time before a Midland 2-6-4T proposal again surfaced. In a scheme for twelve standard designs to meet the needs of the newly formed LMS in 1924, the Horwich drawing office of CME George Hughes outlined a 2-6-4T with 5'6" driving wheels, which was really a tank version of the later 'Crab' moguls of 1926. This too got no further, but the 1914 Derby 2-6-4T plan was revived and eventually appeared with 5'9" driving wheels in 1927, officially attributed to Henry Fowler. The boiler was broadly similar to that on the original 1914 proposal. It was this Fowler 2-6-4T design that was the basis for the NCC Moguls, as we saw in Chapter Eight, though with 6'0" driving wheels.

The NCC 2-6-4T proposal can be traced to as early as 1944. An LMS Derby drawing, by FG Carrier (who later styled the BR standards) and dated 13 March 1944 (Derby 5557B), shows the projected engine with 5'9" driving wheels, and a cab similar to that fitted to the Stanier 2-6-4 tanks. A second drawing (5558B) shows the projected engine with 6'0" driving wheels, which was the NCC standard for the moguls. At this time

Left: No 5 in lined black but with a cheap and nasty NCC on the tank at Belfast in 1947. The gap in the footplating ahead of the cylinders shows the hand of Ivatt in the design. Note the square corners where tank and bunker meet cab. Only the first ten tanks had these. Stress fractures occurred and the last eight tanks had rounded corners, as seen below, while the first ten were modified.
Real Photographs Ltd

Left: No 53 at a standard Queens Quay vantage point in front of the large corrugated iron shed in 1951. The hinge for the tablet catcher inside the cab door is clearly visible.
Real Photographs Ltd

CE Fairburn had just become CME of the LMS, with HG Ivatt as his deputy. As Fairburn was primarily an electrical engineer, most of the steam work was done by Ivatt, who was the real designer of the LMS 2-6-4 tanks usually attributed to his Chief. Ivatt's hand on the NCC tanks can be seen in the gap in the footplating at the front end, and the fitting of a Stanier chimney further forward than on the mogul – possibly to accommodate

the arrangement of the blastpipe and the self-cleaning smokebox. The top feed was also moved a few inches back towards the dome.

Like the Moguls, the 1946 Jeeps had a coupled wheel base of 8'0"+8'6", the last use at Derby of dimensions which had attained almost 'sacred cow' status for six-coupled locomotives since Johnson's time, if not

No 5 came with the tablet catcher below the cab window and low down. The fireman was supposed to swing it out from the cab, but this procedure was awkward and dangerous. No 5 was modified to conform to the later engines' which had a detachable catcher mounted on a bracket inside the cab doorway. Since the tanks were designed to run in either direction their catchers faced both ways.

Real Photographs Ltd

No 5 on the 6.00pm up train at Larne Harbour on 22 April 1948. Local 'timer' RM Arnold (then a young man) is talking to the driver, presumably encouraging him to do a fine run to Belfast. Oval buffers came with the tanks and a screw coupling was fitted at the front. All of the engines received plain unshaded numerals and letters at Derby, in contrast to the attractive NCC shaded version, later perpetuated by the UTA.

W Robb

No 10, the last of the second batch of mogul tanks, at Belfast shed on 28 September 1947.

Real Photographs Ltd

A rear view of No 3 from the 1947 batch, at Belfast shed on 13 April 1948. The levelling pipe between the side tank and the bunker tank can be seen behind the cab footstep.

HC Casserley, courtesy RM Casserley

Left: Although ordered by the LMS, the third batch of 2-6-4Ts, Nos 50–53, were completed at Derby under the auspices of the Railway Executive of BR in April–June 1949. As this was only just after the UTA had taken over, they still had 'NCC' on their side tanks. No 52 is seen brand new at Belfast in 1949.

Author's collection

Below: An unusual combination of mogul tank No 4 with the 'NCC' painted out and 'Jinty' No 19 with the 'NCC' still showing, at Belfast in May 1950. No 19 is shunting carriages. The leading vehicle in No 4's rake is a J10 open third, shortly to become an MED railcar.

GW Sharpe

Kirtley's. The contemporary LMS Fairburn 2-6-4 tanks used a coupled wheel base of only 7'7"+7'9", dimensions perpetuated in the British Railways standard 4MT tanks, which were much more compact engines than the Jeeps. However, the longer coupled wheelbase was retained on the NCC tanks to standardise coupling rods and valve gear with the Moguls. To standardise tyre sizes, the NCC demanded a six foot driving wheel and a three foot pony truck rather than the LMS 5'9" driver and 3'3½" pony. It is interesting to speculate whether the NCC's six foot wheel for a mixed traffic locomotive influenced Stanier in his design of the 'Black Fives' in 1934.

The design can therefore be described as basically a Fowler 2-6-4 tank with cosmetic alterations influenced by Stanier and Ivatt. Although the LMS had been using taper boilers since 1933, the Jeeps remained faithful to the parallel type. Ivatt was also responsible for the rocking grates and self-cleaning smokeboxes which made the tanks unique among Irish engines in being delivered with both features. Engines Nos 54–57 were fitted with exhaust steam injectors. So, although there was no exact equivalent, the Jeeps were instantly recognisable as LMS engines, despite the fact that the NCC continued to use the late Johnson style hand wheel to fasten the smokebox door. The drawings for the tanks were mostly done by the Derby draughtsmen

and, on arrival in Belfast, the engines were assembled under the supervision of J Lofty, a well-known works figure at York Road.

The 1944 Derby drawing also showed a square corner between the cab and tank, which later had to be modified due to cracking. This problem became so acute that engines 50–57 were delivered with a much gentler radius of curve between cab and tank, and nos 1–10 were gradually altered as they passed through the works. Except for Nos 99–104, the mogul boiler had fewer small tubes than the Jeep boiler, and it is interesting that for the last few years of their lives Nos 50 and 51 actually carried mogul boilers. The moguls had a wider, though six inches shorter, firebox.

The first four locomotives, Nos 5–8, arrived in August and September 1946. The 'NCC' on the tank sides, and the numbers on the buffer beams, were rather ugly and unshaded block transfers, and the numberplates,

Above and left: No 6 is delivered at Belfast docks in August 1946. *CP Friel collection*

Below No 7 is about to be lifted to have its bogie refitted after its first overhaul at York Road works in May 1950. The engine has already been repainted and the original plain buffer beam numbers obliterated.

Derek Young collection

also rather crudely cast, lacked the NCC's distinctive curving style. The livery was black, with a quarter inch yellow line round the edge of the bunker, side tanks and cab door, the space between the line and the edge of the tanks being filled in with a maroon colour. In 1949, No 5 was repainted for a short time in an experimental light green livery at the instigation of Harold Houston. (In 1967 Northern Ireland Railways, the successor of the UTA, adopted maroon as a livery, and the late 'Mac' Arnold tried, without success, to persuade the management to paint No 51 in this colour!)

Transporting the engines to Belfast was a story in itself. The boiler and frames weighed 36 tons, and the LMS had only six well wagons capable of carrying this load. When the engines arrived at Heysham in dismantled form a logistical problem arose: the frames were 43'3" long, but the sheer legs of the crane at Heysham were only 36'0" apart. Anticipating this, Derby works made scale models of the tanks and the dockside cranes, and worked out in advance how the engines should be slewed through the gap to be loaded on to the LMS cargo ship *Slieve Bloom*. On arrival at Belfast, Harland & Wolff's 150 ton floating crane was used in the unloading of the tanks.

One problem which Derby probably did not foresee was the question of tablet exchange apparatus. No 5

appeared with the conventional tablet catcher mounted on the cab side, but it soon became apparent that the fireman would have needed the long arms of an ape and the skill of an acrobat to retrieve the tablet from the jaws of the catcher. Moreover, the conventional catcher only worked while the engine was travelling forwards, and the tanks were designed to run as fast bunker first as chimney first. Special tablet exchanging apparatus was therefore designed with a right and left detachable head. This was kept in the back of the cab and manually pushed on to the head of a plug on each side of the engine's cab.

The mechanical dimensions of the WT class can be

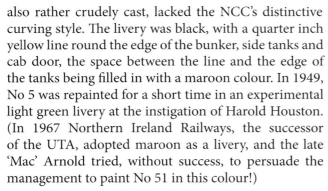

found in Table 27 and the basic chronology of the class in Table 28. Since we are moving into a class of engines well known and carefully recorded throughout their history, it has been possible to assemble a basic history of each engine, and this appears in Table 29.

Six more of the class, Nos 1–4, 9 and 10 appeared in the Spring of 1947 and a further four were ordered in October, for delivery in 1949.

Within a very short time the new tanks were nicknamed 'Jeeps' as a tribute to their ability to go anywhere and do anything required of them. This was a rare case of a nickname becoming a semi-official description, as working notices often referred to 'Mogul and Jeep type engines'.

Like many new engines, the Jeeps were not at first received with open arms, particularly by the Belfast top link. It was claimed that they were not as reliable steamers as the moguls, though experience gradually taught enginemen that they did best with 'little and often' firing rather than the big fire that a mogul thrived on. The blastpipe diameter was found to be ⅛" too broad, and several crews used their own 'jimmies' to sharpen the blast before the matter received proper works attention. On one occasion WS Marshall, the UTA Operating Superintendent, was on the golf course at Whitehead when he saw No 10 bursting through the adjoining Slaughterford Bridge with the down Boat Train, her exhaust like gunfire and the engine furiously blowing off. Next day the fireman was 'on the carpet' to explain his management of the engine. He pleaded that he'd had a failure and that no other engine was available. Marshall had No 10 examined, and found nothing

amiss. But he was well aware that the engine had been 'got at' by her crew, and a good enough railwayman to know that he would never penetrate the wall of silence into which further enquiries were bound to run.

On 1 January 1948 the railways in Britain were nationalised and the LMS became part of British Railways. Thus for 15 months the NCC was part of BR and was known as 'The Railway Executive, NCC'. This name appeared jointly with that of the UTA on the cover of a unique 1948 working timetable covering the whole of the NCC and the BCDR (by then part of the UTA).

During this period work began at Derby on the four Jeeps ordered in October 1947 and an order for a further four was placed in 1948. Materials for all these were ordered and the first four, Nos 50–53, were completed in April to June 1949. However, in the interim, the Railway Executive had sold the NCC to the new Ulster Transport Authority on 1 April 1949. This created an interesting legal problem because, under the terms of the Transport Act, 1947, the Railway Executive could not "construct, manufacture or otherwise produce anything which is not required for the use of their undertaking". An internal memorandum relating to this is reproduced on the next page. Although delivered after the UTA takeover, Nos 50–53 still bore NCC livery. Even though only some of the materials ordered for the construction of Nos 54–57 had been delivered by 1 April 1949, the 'illegal' order was allowed to proceed and the locomotives were delivered in July and August 1950, this time bearing the UTA roundel, but still with block numerals on the buffer beams.

With the takeover by the UTA, the old NCC died on 1 April 1949. The UTA was held in such opprobrium by railwaymen that the name was rarely used. Men who had worked for the companies which it absorbed continued to refer to themselves as 'County Down', 'Great Northern' or 'Northern Counties', and even the UTA's working timetables and notices continued to use the old names to describe their three areas of railway.

From 1950 on, 'Jeeps' entering the shops for their first major overhaul were painted into the new, more elaborately

lined, UTA black livery (see Appendix 2). The tanks dominated Larne line workings and several of them were transferred to the BCDR to assist the Baltics on the Bangor line. The first to arrive was the virtually brand new No 53 in September 1949. She was joined by the now-preserved No 4 in May 1950 and by No 10 in January 1951. In May of the same year No 50 replaced 53 on the Bangor trains. No 4 departed in August 1951, leaving Nos 10 and 50 to hold the fort until No 7 joined them in August 1952. By then, the arrival of the new MED railcars was sounding the death knell of steam on the County Down. No 10 returned to the NCC in August 1952, No 50 in January 1953 and finally No 7 in October.

Meanwhile, No 57 was loaned to the GNRB in May 1952 and operated there until June 1954, when it was replaced by No 7, which stayed until the following year. No 4 joined her in September 1954.

By the mid 1950s the tanks had become much more popular, but the large-scale dieselisation of the NCC main line in early 1957 and the Larne line in 1959, was evidence of management desire to be rid of steam for good, and the future of a class of largely new engines suddenly seemed rather precarious.

The Jeeps in the latter years

The Jeeps came right at the end of the history of Irish locomotive development and, along with the Great Northern VS class 4-4-0s and the two SLNCR 0-6-4 tanks, were the last new conventional designs introduced on an Irish railway. The Great Northern had, of course, built five more engines each of the well-tried U class 4-4-0s and UG class 0-6-0s, but it could be said that the tanks were the most versatile of all the engines built after 1945 and certainly they outlasted all the rest.

The writing was on the wall, though,

APPENDIX
(vide R.E. Min.2504)

MEMORANDUM TO THE BRITISH TRANSPORT COMMISSION (Enclosure to R.E. letter to B.T.C. dated 17/5/49.)

CONSTRUCTION OF LOCOMOTIVES FOR
NORTHERN COUNTIES COMMITTEE

1.   It was the practice of the former L.M.S. Company to build locomotives for the Northern Counties Committee, ten 2-6-4 (5'3" gauge) passenger tank locomotives having been constructed in L.M.S. workshops for the Committee during 1946 and 1947.

2.   In October, 1947 the L.M.S. Board approved a proposal to construct a further four 2-6-4 passenger tank locomotives for the Northern Counties Committee at an estimated cost of £39,640; two of these have been completed and the work in connection with the other two is well in hand. In 1948, it was agreed to build an additional four locomotives of the same type in 1950. It is anticipated that the cost will be approximately the same as for the four engines which are being built this year, namely £39,640. All the material for the four locomotives it is proposed to build in 1950 has been ordered and some of it has been received.

3.   In view of the fact that the undertaking of the Northern Counties Committee was transferred to the Ulster Transport Authority as from 1st April, 1949, it is appreciated that it is not permissible for the Executive to manufacture rolling stock for them as this would contravene the express provision of s.2 of the Transport Act 1947, namely that the Commission shall not "construct, manufacture or otherwise produce anything which is not required for the use of their undertaking". On the other hand, as the work on the construction of four locomotives has either been completed or is well in hand and a proportion of the material ordered for the four to be built in 1950 has already been received, the Executive's Legal Adviser and Solicitor has expressed the opinion that there would be no objection to the programme being completed.

4.   It is recommended, therefore, that the proposal to build eight locomotives at an estimated cost of £79,280 be approved, on the understanding (a) that the Ulster Transport Authority will pay the actual cost of constructing the locomotives and, in addition, costs of dismantling, preparation for despatch to Heysham, assistance in re-assembly at Belfast, carriage by rail and shipping charges and (b) that no further commitments will be undertaken.

(Sgd.) E. J. MISSENDEN.

May, 1949.

The internal BR memorandum about the construction of the Jeeps. *Courtesy Plil Atkins*

Right: Several of the 2-6-4Ts went to Queen's Quay for a spell. No 53 was there from September 1949 to May 1951 and is seen in front of the shed in 1950.

Real Photographs Ltd

Opposite top: In this view No 53 is positioned well to show the old BCDR shed at Queen's Quay.

Real Photographs Ltd

Below: No 8 at Larne Harbour making up the 'perishable' train about 1955. The man with the rope is controlling the level crossing in front of No 8. The car is a 1936 Brighton-registered Wolseley Wasp with a missing bumper.

A Donaldson

Right: The NCC was not noted for its scenery, but the stretch between Castlerock and Magilligan was the exception. No 3 is seen here passing Umbra gates with the 2.40pm ex-Belfast, as it runs along the Downhill cliffs towards Bellarena about 1957. At the gates a milk lorry awaits its turn.

A Donaldson

Opposite bottom: No 10 had an extended spell on the BCDR from January 1951 to January 1954, working on the Bangor line. She is pictured here at Queen's Quay with a long rake of antiquated stock. The County Down drivers liked the tanks and Paddy Fitzpatrick reached 70mph with one on an up train through Holywood.

Author's collection

Above: How to save time and give the civil engineer a heart attack! Five engines, Nos 54, 7, 50, 53 and 97, with a combined weight of 458 tons, are despatched from Portrush to Coleraine shed in June 1959.

WA Scott

Centre: In the summer, it was common practice to store excursion carriages down the steeply graded harbour branch at Portrush. On 28 May 1966, two engines have failed to extract a long rake and No 10 is now going down to add her muscle to the job.

Author

Right: No 10 brings a special from Londonderry through Coleraine in 1967. The fireman is handing over the tablet to the Coleraine signalman. On the extreme left is the mile-long Coleraine harbour branch, then two sidings before the junction with the Portrush branch. The water tower and ash pit are on the first siding.

Author

for steam in Northern Ireland. Before the class was complete, the UTA had begun a large scale programme of railway closures. By 1959 the NCC main line and Larne line had been substantially dieselised, and a few moguls and Jeeps were routinely put into winter storage at the end of the busy summer season. For the moguls, all this was the first tolling of the funeral bell; for the Jeeps, it was the beginning of a dispersion to the Great Northern and, in retrospect, a quite lengthy extension to their lives.

The first transfer to the GNRI was No 51 in September 1958, followed quickly by Nos 53 and 56. They were instantly popular. There were several reasons why the tanks received a much warmer reception from Great Northern men than the moguls. The engines sent to the GNRI were generally in better condition, rode more smoothly than the moguls, had a much warmer and more comfortable cab and would run fast and freely on favourable gradients.

Maintenance was generally good, and reference to Tables 27 and 29 will show that

Above left: With shaded numerals, No 53 is seen at Balmoral on the GNRI main line on 2 July 1953, on the occasion of the 'Coronation Youth Gathering' in the King's Hall, during the Royal visit of that year. Twenty trains were dispatched from Balmoral with the returning participants, mostly uniformed youth organisations like the Scouts. No 53 has the 5.55pm return special for Kilrea. The train includes all four ex-BCDR clerestory vehicles, two of which are now preserved at Downpatrick. This picture bears remarkable similarity to the one taken at the same location in 1937, on page 113.

William Robb

Left: On an early visit to the GNRI in 1953, still wearing her original unshaded numerals on the front bufferbeam, No 57 worked the 5.00pm Londonderry train to Portadown and the photograph shows her about to hand over to GNRI S2 class No 190.

A Donaldson

Right: Two NCC engines at Great Victoria Street station in 1953 – 2-6-4T No 7 at Platform 3, in company with U2 class 4-4-0 No 72 at Platform 2. Visible through the Boyne bridge is GNRI U class 4-4-0 No 200 at Platform 1. No 7 has been to the paint shop and got a proper shaded number rather than the cheap one she came with. Mail bags are being loaded into the train.

A Donaldson

Dundalk takes on an NCC appearance about 1961! No 1 has come off an up train, to hand over to a CIE diesel, and heads off to the turntable past 2-6-0 No 97 which is building up a large fire to face the Wellington bank on the way back north.

Author

there was a useful shopping and reboilering programme in the early 1960s. The engines which received mogul boilers probably received the standard number of tubes, as fireboxes and tube-plates were replaced. The mogul regulator valve was reputed to have survived, but few comments were ever made by drivers about any differences they noted in the port openings.

The Jeeps were also used on work which was well within their capacity. By the time they came in numbers, the GNRI was split between the UTA and CIE, and there was very little through southbound steam working beyond Dundalk. Based mainly at Portadown, the Jeeps in the last years of Great Northern steam did some of their best work on moderately loaded semi-fast trains between Belfast and Dundalk, and on local services from Belfast. Until the very end of steam they were kept off the 'Derry Road', thus avoiding the punishment that the Moguls took with heavy goods trains on this difficult line.

Portadown drivers in particular took to them with enthusiasm, and the standard of work they produced was always good and sometimes excellent. Speeds in the seventies were not uncommon and No 55, with four bogies on the 6.30pm ex-Dublin, gave Mac Arnold a run from Lurgan to Lisburn in 13'22", start to stop for 12½ miles, with max of 81mph. With a much heavier

load, one of the best 1960s performances was done by the same engine on the same train with an eight-bogie load. A time of 25'36" from Dundalk to Goraghwood was well up to the best pre-war standards. Finally, the tanks excelled on the difficult 20 minute timing from Goraghwood to Portadown which, although only 15¾ miles, had three severe permanent speed restrictions at Poyntzpass, Scarva and Portadown Junction. Two of the best of many good efforts was a time of 18'01" by No 56 with a light train, and an excellent 20'16" by No 57 with a sizeable 310 tons. Possibly the most astounding effort over the Wellington bank was on a train which, to my knowledge, included no timers among its passengers. The late Drew Donaldson was out photographing one day in 1960, and to his amazement found himself taking a shot of 53 running down the Wellington bank with no less than eleven bogies behind her.

It was, though, a courageous driver who took No 53 over the bank without assistance that day, for the general perception on the Great Northern was that the Jeeps were better runners than steamers. Whereas NCC drivers would customarily work a Jeep with about half the main valve and about 25% cut off, Great Northern men would use a less generous regulator opening and a longer cut off. Great Northern men took some time to get used to the Jeeps' injectors (Was it a coincidence that some of the best Jeep work on the Great Northern was often done with one of the four engines fitted with exhaust injectors?) and to the combination brake, which NCC men used so skilfully in making very fast approaches to station stops. A further problem was braking power, for the Jeeps performed genuine mixed traffic work, typically hauling a heavy freight over the bank to Dundalk and returning with a moderately loaded, and quite tightly timed, passenger train.

The closure of the Derry Road in February 1965 led to the closure of Portadown shed, the end of steam for all but special and relief trains, and the scrapping of all but a handful of the last GNRI passenger engines.

For steam, this was a rather sad period of limbo on the Great Northern, though there was talk of replacing Adelaide with a smaller and more modern shed at Great Victoria Street. A turntable (from Ballymena) was indeed installed in the vicinity of the old Ulster Railway shed yard, but no further steps were taken. In default of any bigger engines after the withdrawal of No 207 in September 1965, tanks were called on

to work fairly heavily loaded trains all the way from Belfast to Dublin, a distance of 112½ miles, at a time when coal was no longer available at Dublin following the elimination of steam on CIÉ in 1963.

The first attempt to solve this problem involved the conversion of a small mogul tender to be hauled by a tank, with an auxiliary supply of coal and water. Engines 53 and 55, and later 51 and 56, were modified to work with this tender but, since water does not generally flow uphill, the tender was found to be less than a success on a line with steep banks. Nor did firemen appreciate having to spend time at Dublin shovelling coal from the tender into the restricted bunker area of their engine. Water was available at Dundalk and Drogheda, but the coal problem required a more practical solution.

An eminently sensible idea, pioneered on No 54 during her last overhaul in 1965, was to fit a bunker extension ('crib') enabling the engine to carry another 1½ tons of coal. A second set was sent to Adelaide and fitted to No 53, but she became a temperamental performer and they were soon transferred to the more

Right: The signalman on his balcony at Knockmore Junction hands over the staff to No 53 on 28 May 1960. No 53 was on a train known as the 'Fred', so called because this annual excursion to Portrush was organised by enthusiast Fred Graham for the Sunday school children from Windsor Gospel Hall and often had specially requested motive power.

Author

Below: No 55 is working hard as she climbs towards Adavoyle summit. Crews were always glad to pass this point as the gradient eased here before the final rise to post 65¼.

A Donaldson

Above: By 1964 the tanks were permitted over the GNRI Derry Road but under severe speed restrictions. No 57 has penetrated to Omagh and is working an up train to Belfast.

Des FitzGerald

Left: No 50 on an up goods near Adavoyle between Dundalk and Goraghwood. Using the tanks on goods trains caused some trepidation among GNRI crews, as they did not have the braking power of a tender engine like a GNRI 'Big D' 0-6-0 or an NCC mogul. Note the immaculate track and verges.

A Donaldson

Opposite: The stone train contract between 1966 and 1970 prolonged the life of the mogul tanks as they hauled 7,600 trains of 1,000 tons, Nos 53 and another are seen at Slaughterford bridge, between Larne and Whitehead with an up stone train about 1967. *Author*

reliable No 56. A third set was made for No 55. In an emergency, 'non cribbed' engines still worked to Dublin, and passengers would be intrigued to see the front guard's van filled with sacks of coal! No 4, the only single-numbered engine to carry them, was fitted with cribs taken off No 56 in 1969 as part of her preparation for the RPSI's 'Brian Boru' rail tour to Cork on 27–28 April 1969 – the first time an engine of this class had ever appeared so far south. Reference will be made later to the Magheramorne Spoil Contract, for which the regular engines received additional sets of 'cribs' to eliminate the need to coal the engine during a shift. Nos 4, 50, 51, 53, 54, 55 and 56 were all so fitted for these trains, and No 4 still carries hers in preservation. I am indebted to Irwin Pryce for the table of variations between individual engines of the class which appears as Table 29 (pages 136–38). It certainly gives the lie to any idea that all the tanks looked exactly the same!

By 1966 it seemed that the end had come. There was the late spring flurry of Sunday School excursions, and the heavy summer programme of steam-hauled relief trains on both the GNRI and the NCC, but very little in the way of steam work all the year round. Nos 2, 7, 8 and 52 went into storage and what became permanent withdrawal. Worse still for steam enthusiasts, was the announcement of a new generation of diesel electric railcars, which eventually did finish steam off altogether.

But an Indian summer was coming for steam, beginning in early 1966 when the ageing fleet of MED and MPD railcars began to break down with increasing frequency. To the embarrassment of management, and the delight of steam enthusiasts, a significant number of daily Larne line services reverted to steam haulage to facilitate a major diesel repair programme. On a summer Saturday the 8.35am to Portrush, the 8.50am to Derry, and often the 3.05pm to Portrush, would be steam, as well as at least one of the Larne line boat trains, especially during the 'Glasgow Fair' period. Right down to 1968, the busiest day of the year was Apprentice Boys day in Derry in mid-August, when six specials, and sometimes a steam-substituted regular train as well, would bring the roofless shed at Waterside back to life with a vengeance. There were, too, very occasional days when major steam spectaculars occurred: one thinks of the last Saturdays in August 1965 and August 1967 when Royal Black Preceptory Demonstrations at Lurgan and Carrickfergus led to the running of over eighty separate loaded and empty carriage movements in the course of a single day.

The mood at York Road shed brightened even more when it was announced that the railway had won a major contract to shift stone filling from Magheramorne quarry to the shore of Belfast Lough, where a major new motorway was being built on reclaimed land. This undoubtedly hastened the end of what little steam survived on the Great Northern. In November 1966 Adelaide shed closed, and for the first time since 1959 all the workable Jeeps were back on the NCC where York Road shed rapidly became fuller than it had been for years. The best engines were taken into the works to be 'soled and heeled', and Inspector Frank Dunlop – always a good manager and psychologist – persuaded the management to revert to an old practice and allocate engines to regular crews on the Stone Train link. This had an immediate pay-off, not only in morale but also in the state of the engines, some of which were kept highly polished externally, and with cabs that glittered like jewellers shops.

So, while steam finished on British Rail in August 1968, eight hundred ton trains of spoil, topped and tailed by Jeeps, were storming up Mount bank six days a week, in addition to whatever passenger trains needed a steam engine as well. In the summer of 1969, the last in which steam appeared regularly on the main line, such haulage feats took place out of Belfast as eleven-bogie trains worked to and from Portrush on several occasions without assistance.

Exciting as this was to watch and photograph, the engines were, sadly, being worked to death by a management who needed them only to complete the Magheramorne contract. By May 1970 that contract was nearly complete, the last of the engines were mechanically far through, and the completion of the M2 motorway involved the annihilation of the York Road shed complex. Years previously, an anonymous UTA executive had smugly predicted; "There won't be a puff of steam about the place after 1960". The final years of sterling service by the 2-6-4 tanks had proved him wrong by a full decade.

Performance on the NCC

Fortunately, the Jeeps did most of their work during a period when there were train timers in plenty to record their best work for posterity, and even in their last run-down days they were capable of performance which would not have shamed the NCC in the 1930s. The first batch of Jeeps were largely confined to the Larne line, where five Scotch engines were replaced by Nos 2 to 5 of the new tanks. Each had its own pair of regular crews, except for No 2, which worked a seventeen hour turn of duty that required three regular crews. In the longer term Nos 1, 8, 9 and 51 became the regular Larne engines, but it took longer for the Jeeps to be fully accepted on the main line, and it was some time before they were regularly based at Coleraine or Derry. On the NCC, the Great Northern, and the Bangor line, they eventually set new standards of performance, and in preservation No 4 has performed with great distinction on each of the main lines out of Dublin over the last thirty years, arousing admiration everywhere she has gone.

First, then, to the Larne line. Run 1 is a fine effort by No 55, then only a few weeks old, on the 5.50pm Boat Train. Run 2 was on a special train of three North Atlantic coaches and bogie van which took the RBAI Railway Society non-stop to Larne Town and produced probably the highest speed ever done down Mount bank. Run 3 is a scorching effort on the 5.20pm stopping train, producing times to Carrick and Whitehead that can rarely have been bettered by steam. Runs 4 and 5 feature two excellent efforts over the taxing level road between Coleraine and Derry.

Runs 6–8 along with No 91's epic run on the 'North

In NIR days some of the Jeeps were used on RPSI railtours. While in Dublin for the *Slieve Cualann* railtour on 4 May 1968, No 56 went to Enfield for gauging purposes in connection with a planned two-day tour on former Midland Great Western lines, scheduled for September. She is seen here at Clonsilla. Gauging was approached very simply in the 1960s. Lead strips were placed on the widest parts of the engine and then examined for scratches.

Dave Murray

Atlantic Express' represent some of the finest work done in the up direction on the NCC. In February 1958 the main line was dieselised, and the expresses given a 110 minute schedule from Derry to Belfast that proved impossible even for the MPD units. Several times in one week, however, one of the diesel sets had to be steam substituted, and engines 99, 54 and 2 were involved. On the last day of this eventful week, Driver John McAllister of Coleraine produced the superb effort in Run 6, and without taking water either!

Few runs on any railway merit the description 'best ever', but Run 7 must be in this category. Signal stops at Macfin and Ballyboyland annoyed the mild-mannered driver Alan Robinson, and at Ballymena he announced that he would be "right time Belfast". The restart from here was probably the best ever start to Kellswater, and the best ever climb from Antrim to Kingsbog. Antrim to Kingsbog was one of the longest block sections on the NCC at this time, and the train was checked at the Antrim advance starter, and again at Mossley as he commenced a whirlwind run down into Belfast.

Finally, in Log 8 is another run with No 4, driven by Tom Crymble, which probably put out a greater DBHP than either of the other two, as he topped the bank at 59mph with a 60 tons heavier train, equivalent to 900dbhp. The train was taken deliberately easily out of Ballymena, in the knowledge that an MED set was just ahead on a local to Belfast, and there was indeed a signal check at Antrim. But Fireman Arthur McMenamin certainly earned his wages as No 4 lifted 230 tons up to Kingsbog in dramatic fashion. Right to the end the Jeeps were capable of good hill-climbing. Indeed with only months to go before the final end of steam, No 55 in June 1969 managed to hold a minimum speed of 40mph at Kingsbog with a mammoth eleven bogie return Sunday School excursion from Portrush.

Finally, Runs 9 and 10 show two contrasting examples of Jeep working on down trains. Run 9 dates

from the earliest days of the class, and records the fastest speed ever to have been done by a Jeep. Run 10 shows that, even in their last days, a tank engine could handle a big train economically. Driver McAleese produced a fine effort, and made one of the last non-stop runs from Kingsbog to Portrush.

As previously mentioned, No 4 has survived to the present day in the ownership of the Railway Preservation Society of Ireland. Her day-to-day maintenance has given a rare opportunity for railway enthusiasts to do the work which in the 1950s and 1960s had to be carried out by railwaymen, and to conclude this chapter I am indebted to Irwin Pryce for some rarely-noted aspects of Jeep design and performance.

The mogul tanks were a product of their time, and the finish was poor where it was not important: for example, tool marks on the coupling rods. Tanks and bunkers were a mixture of riveted and welded construction, and frame stretchers were made of plate where formerly there would have been a substantial casting. This saved weight but led to corrosion under the cab and at other places. Possibly due to the use of inferior steel, problems occurred.

Cylinders were attached to frames with fitted bolts. These eventually worked loose due to the constant 'shouldering' effect of any engine with two outside cylinders. Attempts were made in the works to weld around the cylinders, but the welds invariably cracked. Bolt holes needed to be reamed out and bolts a size larger made – moguls had similar problems. Frame fractures were more common than on moguls – a problem never encountered on the GNRI three-cylinder engines, but common enough on the CIE Woolwich moguls.

Axle boxes were lubricated solely by pad and oil box underneath. This was excellent in theory, giving an unbroken bearing surface, and oil fed in at the crown of the box where pressure was greatest. On the other hand, oiling twelve boxes had to be done from underneath, and they collected water when the engine was at a water column or having a boiler washout. The boxes had drain plugs which had to be used if hot boxes were to be avoided.

Injectors were noisy and could be wasteful, but they were easy to adjust. An experienced fireman could work the steam valve with his hand, and the water valve (which was at cab floor level) with his foot.

Multiple valve rings were a good thing to avoid steam leakage.

A rocking grate, hopper ashpan and self-cleaning smoke box all helped disposal. Fire bars were expensive to replace, and very heavy! The self-cleaning screens in the smokebox were very heavy, and of course became filthy, and they made access difficult for tube cleaning and boiler wash outs. The self-cleaning smokebox may have affected steaming ability, which in the early days was inferior to the moguls.

The boiler was too small, with a short glass and a small water space. But with good coal, clean tubes and sensible handling there were no problems.

There was no proper shovelling plate and a heap of coal would accumulate on the footplate.

Despite Major Pope's dislike of turning, failure to turn the engines could lead to uneven tyre wear.

York Road expanded big tubes only, before fitting ferrules. Derby procedure was to thread tube hole and tube and then expand them also.

Table 27: Dimensions of the WT class 2-6-4Ts

Type	2-6-4T
Cylinders	19"x 26"
Piston valves	9 inches
Coupled wheels	6' 0"
Bogie and pony wheels	3' 1"
Wheel base	9' 0" + 8' 0" + 8' 6" + 5' 9" + 6' 6"
Boiler length	10' 10½"
diameter	4' 6⅞"
tubes	138 x 1¾"
Superheater tubes	21 x 5⅛"
Heating surface	1416¾ sq ft
Firebox case	7' 6" x 4'5¼"
Grate area	25 sq ft
Boiler pressure	200 psi
Tractive effort	22,160 lbs
Weight	87 tons
Water capacity	2500 gallons
Coal capacity	3½ tons (4½ tons with extended bunker)

Notes: Nos 54–57 had exhaust steam injectors.

Table 28: Chronology of the WT class 2-6-4Ts, built at Derby

No	Deliv	Cost	Order No	Boiler changes	Last overhaul	Last used	Ran on Bangor line	Scrapped
1	4/1947	£10,320	669	never	1962	8/1966		3/1969
2	5/1947	£10,320	669	never	1963	6/1965		3/1969
3	5/1947	£10,320	669	1/1964 (ex-5)	1964	3/1969		6/1970
4	5/1947	£10,320	669	1/1965 (ex-51)	1965	10/1970	7/1950–8/1951	preserved
5	8/1946	£11,013	1674	11/1963 (ex-10)	1963	3/1970		10/1970
6	9/1946	£11,013	1674	6/1965 (ex-4)	1965	4/1970		9/1970
7	9/1946	£11,013	1674	never	1962	4/1965	8/1952–10/1953	3/1969
8	9/1946	£11,013	1674	never	1962	10/1965		3/1969
9	6/1947	£10,320	669?	never	1962	2/1967		3/1969
10	6/1947	£10,320	669?	7/1963 (ex-50)	1963	10/1969	1/1951–1/1954	6/1970
50	4/1949	£11,347	3283	3/1963 (ex-99)	1963	1969	5/1951–1/1953	10/1970
51	5/1949	£11,347	3283	8/1968 (ex-98)	1964	1/1970		2/1971
52	5/1949	£11,347	3283	never	1963	1/1966		3/1969
53	6/1949	£11,347	3283	5/1964 (ex-3)	1964	5/1970	9/1949–5/1951	6/1971
54	7/1950	£11,717	4332	4/1965 (ex-56)	1965	4/1967		2/1970
55	7/1950	£11,717	4332	7/1964 (ex-53)	1964	3/1970		10/1970
56	8/1950	£11,717	4332	10/1964 (ex-55)	1964	4/1969		10/1970
57	8/1950	£11,717	4332	never	1963	8/1966		3/1969

Table 29: History of each locomotive in the WT class

No 1

In traffic April 1947. First allocated to Belfast top link, then to Larne. Last overhaul 1962. Transferred to GNRI area 1964. Returned to NCC after closure of GNRI Derry Road in February 1965. Withdrawn August 1966, having spent her last days as Belfast station pilot. Scrapped March 1969.

No 2

In traffic May 1947. First allocated to Larne, then to Belfast. First worked on the GNR from July 1962 to February 1963, when she was severely damaged in a collision with AEC railcars 6 and 7 at Lisburn in February 1963. Her damaged smokebox and pony truck were repaired and she then returned to the GNR until February 1965. She was the first of the class to be withdrawn, in June 1965. Scrapped March 1969.

No 3

In traffic May 1947. First allocated to Larne. After the arrival of No 51 she became a spare engine, and never left the NCC until 1968. Last overhaul and reboilering in January 1964, with a reconditioned boiler from No 5, with a new firebox. Latterly in bad condition with cracked frames and used as the regular ballast engine. Used regularly on GNR area ballast work in 1968, still based at York Road. Withdrawn in March 1969 and scrapped June 1970.

No 4

In traffic May 1947. First allocated to Larne, but transferred to the Bangor line from 17 May 1950 to 16 August 1951. Also loaned to the GNRB from 29 September 1954 to 27 December 1955, following the collapse of the Tolka Bridge, and shedded at Amiens Street, later at Drogheda. Otherwise she spent most of her working life at York Road. She had her last heavy overhaul in January 1965, with a reconditioned boiler from No 51 and a new firebox. One of only two of the second batch (the other being No 10) to be regularly used on the Magheramorne Stone contract after 1966. Fitted with an extended bunker ('Crib') to increase coal capacity when working stone trains. Worked the last steam passenger train on NIR on 31 March 1970 and, with No 53, the final stone train on 2 May 1970. Last of the class to be steamed in company service, on 22 October 1970. Bought by the RPSI in June 1971. As the only engine of the class to be preserved, she has travelled virtually the entire Irish railway network in RPSI ownership.

No 5

First of the class in traffic, 8 August 1946. First allocated to Larne, then to Derry, but spent most of her working life at York Road. Painted in an experimental green livery in August 1948. Last overhauled in November 1963, when she received No 10's reconditioned boiler. One of only two of the original batch (the other being No 6) to be regularly used on the Magheramorne Stone contract after 1966. Used regularly on GNRI ballast and stock transfer work in 1968–69. Worked one of the last revenue-earning NIR steam passenger trains on Easter Monday, 30 March 1970, her last day in service. Mileage on withdrawal over 800,000, the highest of the class. Scrapped October 1970.

No 6

In traffic August 1946. Worked virtually all its life on the NCC, apart from a year on the GNR section from September 1960 to September 1961. She was the last Jeep to be overhauled, in June 1965, when she received the reconditioned boiler from No 54. Regularly used on the Magheramorne Stone contract after 1966 until deterioration of her mechanical condition. Saw sporadic use in 1969 and 1970. Withdrawn April 1970 and scrapped in September that year.

No 7

In traffic September 1946. Painted in an experimental unlined black from August 1948 until her next shopping in 1950. Used on the Bangor line from 11 August 1952 to 26 October 1953, and transferred to the GNRB from 17 June 1954 to 28 November 1955. Following the dieselisation of the NCC main line, she spent long periods as the Coleraine spare engine, being lit up as required for ballast work. Last overhauled in 1962.

Transferred to GNR area in December 1964 and after the closure of GNR Derry Road in February 1965, was stored first at Portadown, then at Adelaide until April 1965. After brief use at York Road, it lay out of use from December 1965 and was stored at Ballymena after May 1966. Scrapped in March 1969.

No 8

In traffic September 1946, Allocated first to Coleraine, then to Larne and finally to Belfast. Last overhaul 1962. Transferred to GNR area in mid-1964, and stored at Portadown after the end of regular steam until May 1965. Did some work on the NCC in the summer of 1965 before withdrawal in October. Scrapped in March 1969.

No 9

In traffic June 1947. First allocated to Belfast, then to Larne. Following shopping in 1958, it spent a period on the GNR section, returning to the NCC in July 1959. Last overhaul 1962. It had a further visit to the GNR from Autumn 1964 to July 1965, though stored out of use at Portadown after the closure of the GNRI Derry Road in February 1965. Returned to NCC in July 1965 and used around York Road and on stone trains until withdrawal in February 1967. It was then cannibalised to repair No 56. (See below). Scrapped March 1969.

No 10

In traffic June 1947. Used on the Bangor line for at time from 28 January 1948, and then again from 11 January 1951 to 8 August 1952. Reboilered in July 1963 with the reconditioned boiler off No 50 and a new firebox. Employed regularly on stone trains after 1966. Always a York Road engine but appeared regularly on GNR area ballast work in 1968 while shedded at York Road. It even worked a passenger excursion on 31 May 1969, on the first leg of a run (as far as Antrim) from Portadown to Portrush. Withdrawn October 1969 and scrapped in June 1970.

No 50

Into traffic April 1949. First allocated to Larne and then Belfast. Transferred to the Bangor line from 25 May 1951 to 29 January 1953. Worked regularly on the GNR area after 1961. Reboilered with the boiler off mogul No 99, and a new firebox, in March 1963. Stored at Portadown after the closure of the GNR Derry Road in February 1965 but returned to service in the summer, running briefly at York Road before returning to Adelaide until July 1966. A regular performer on the Magheramorne spoil contract after 1966, and fitted with an extended bunker ('Crib') to increase coal capacity when working stone trains. Withdrawn January 1970 and scrapped that October.

No 51

Into traffic May 1949. First allocated to Larne. It operated on the GNR area from mid-1958, except for a short transfer back to the NCC for the summer of 1961. In August 1964 it received the boiler from mogul No 98, with a new firebox and was modified in 1965 to work to Dublin with an auxiliary tender, though the fitting was never used (see below). Returned to York Road following the closure of Adelaide shed in November 1966. A regular performer on the Magheramorne spoil contract after 1966, it was fitted with an extended bunker ('Crib') to increase coal capacity when working stone trains. It was one of the last two NIR steam locomotives to remain in traffic, being used for occasional shunting after the end of the spoil contract, until 16 October 1970 (No 4 lasted another week). She was scrapped in February 1971.

No 52

Into traffic May 1949, and allocated to Larne for most of the 1950s. Based on the NCC for her entire life. Last overhaul 1963. Seen once at Adelaide on a ballast working in April 1964 – possibly her sole visit to the GNRI. Withdrawn January 1966 and scrapped in March 1969.

No 53

Into traffic June 1949. After running in, she was allocated to Queens Quay from 6 September 1949 to 16 May 1951, the first 'Jeep' to operate on the Bangor line, apart from No 10's visit in 1948. On return to the NCC she worked mostly on the main line until dieselisation of the principal services in 1958.

Worked on the GNR area from July 1959 until June 1966. Modified in 1964 to work with an auxiliary tender for Belfast–Dublin specials. In May 1964 she had a heavy overhaul, receiving the reconditioned boiler from No 3 and a new firebox. She was fitted with 'cribs' in 1965, but these were transferred to No 56, reckoned a more reliable engine for Dublin specials.

She was a regular performer on the Magheramorne spoil contract after 1966, and fitted (for the second time in her life) with an extended bunker to increase coal capacity when working stone trains. With No 4, she worked the last stone train on 2 May 1970, her last duty. Stored at Carrickfergus for some time before scrapping in June 1971, the last 'Jeep' to be cut up.

No 54

Into traffic July 1950 and allocated first to Belfast and then to Derry, until dieselisation of the principal services in 1958. Worked on the GNR area from September 1961 until the closure of Adelaide shed in November 1966. When shopped in April 1965 she received the reconditioned boiler from No 56 and a new firebox. The following month she became the first engine to be fitted with an extended bunker for Belfast–Dublin specials. She was stopped in April 1967, following discovery of defects her new Swindon-built firebox, and was scrapped in February 1970.

No 55

Into traffic July 1950 and allocated to Derry until dieselisation of the principal services in 1958. Worked regularly on the GNR area from the spring of 1960. Reboilered in July 1964 with the reconditioned boiler from No 53 and a new firebox. Modified in 1965 to work with auxiliary tender for Belfast–Dublin specials. Fitted with extended bunker in May 1965 also for these specials. Returned to the NCC following the closure of Adelaide shed in November 1966. A regular performer on the Magheramorne spoil contract after 1966, and retained the extended bunker to increase coal capacity when working stone trains. Withdrawn March 1970 and scrapped that October.

No 56

Into traffic August 1950 and allocated to Belfast until dieselisation of the principal services in 1958. Credited with the highest speed ever done by a member of the class – 86mph in August 1950. At her last overhaul in October

1964, she received a reconditioned boiler from No 55, with a new firebox.

Worked regularly on the GNRI area from July 1959, being modified in 1965 to work with an auxiliary tender for Belfast-Dublin specials, and then the 'cribs' from No 53. Damaged in a collision during the lifting of the GNRI 'Derry Road' in July 1966 and stored at Adelaide. She returned to the NCC following the closure of Adelaide shed in November 1966 and repaired in 1967, using the pony truck and smokebox door off No 9.

A regular performer on the Magheramorne spoil contract after 1966, and retained the extended bunker to increase coal capacity when working stone trains. Withdrawn following damage to a cylinder on 29 April 1969.

No 57

Into traffic August 1950. Possibly unique in the class as seeming to have run more miles on the GNR area than on the NCC. Loaned to the GNRB from 15 May 1952 to 17 May 1954 and possibly on another occasion in the 1950s. Allocated to the GNR area from 1960 to May 1966. She was the only one of the last batch not to receive a reconditioned boiler, though overhauled in 1963. Stored at Adelaide after February 1965 before hauling No 7 to Ballymena for storage. She returned to York Road in August 1966, but was used only for shunting, principally on 13 August. She was last steamed on 27 August and scrapped in March 1969, with 465,716 miles to her credit, the lowest of the class.

Run	1		2		3	
Date	27 July 1950		28 Sept 1963		25 July 1958	
Train	5.50pm ex-Belfast		9.55am spl ex-Belfast		5.20pm ex-Belfast	
Loco	**55**		**4**		**6**	
Load	270 tons		125 tons		200 tons	
Belfast	**00.00**		**00.00**		**00.00**	
Whitehouse	15.15	47 *signals*	05.47	58	05.23	57
Whiteabbey	08.45	43	06.58	57	06.32	56
Jordanstown	10.13	48	06.06	57	07.41	57
Greenisland	11.48	48	09.26	59	09.03	56
Trooperslane	12.47	53	10.33	70	10.12	
Mount			11.19	72	11.00	69
Carrickfergus	16.37	58	12.06	61	**12.12**	
Downshire Park			12.52	65/66	02.34	58
Kilroot	16.37	67	13.55	64	03.04	63
Whitehead	19.49	54	17.22	43	**06.28**	
Post 16			18.54	52		
Ballycarry	21.45	67	19.31	60		
Magheramorne	25.15	*signals*	23.30	32 *signals*		
Glynn	27.11	60	25.38	63		
Larne Town	29.07		**28.29**			
Larne Harbour	**32.00**					

Run	4		5	
Date	?		5 September 1964	
Train	?		8.35am ex-Belfast	
Loco	**55**		**5**	
Load	150tons		180 tons	
Coleraine	**00.00**		**00.00**	
Post 66	06.25	64	05.52	63
Castlerock	**07.35**		**07.45**	
Downhill	03.20	48/53	02.48	56
Magilligan	06.10	59	05.47	65/70
Post 74	08.13	61		
Bellarena	09.10	61	08.32	73
Post 76	10.13	61	10.22	73
Post 78	11.13	65	11.11	74
Limavady Junction	**14.15**		**13.15**	
Post 82	04.07	52	04.31	
Post 84	06.20	55	05.47	61
Post 86	08.21	59/61	07.47	66
Eglinton	10.15		09.21	69
Post 90	13.29	21 *tsr*		
Culmore	14.15		12.58	22 *tsr*
Lisahally	15.25	*signals*		
Post 93	04.22	50	17.47	60
Londonderry	**08.10**		**22.09**	

Run		6		7		8	
Year		15 Feb 1958		4 Sept 1965		17 Aug 1968	
Train		2.20pm ex-Londonderry		3.35pm ex-Portrush		Special ex-Ballymena	
Loco		**2**		**4**		**4**	
Load		160 tons		170 tons		230 tons	
Miles	(from Portrush)						
06.0	**Coleraine**	**00.00**		**00.00**	60 max		
10.6	Macfin	06.28	57/65	07.27	*signal stop*		
14.3	**Ballymoney**	**10.19**		**05.30**	61 max		
	Ballyboyland			*06.11/08.54*	*signals*		
17.7	Post 50	05.53	37 *tsr*				
21.6	Dunloy	10.24	62	14.53	69		
24.3	Killagan	12.48	70	17.09	75		
26.5	Glarryford	*15.31/21.32*	*signals*	18.57	71		
	Post 39 (psr Dunminning curves)	25.25	58	21.00	60		
31.3	Cullybackey	27.47	70	23.14	75		
34.2	**Ballymena**	**30.43**		**26.49**		**00.00**	
38.4	Kellswater	05.13	72/75	04.49	80	06.41	61
42.8	Cookstown Junction	08.43	70/75	08.17	70/75	11.00	65/66
45.9	**Antrim**	**11.37**	57 *signals*	**11.34**		**14.55**	25 *signals*
47.9	Muckamore	13.28	65	04.21	*signals*	18.08	58
49.3	Dunadry	14.43	68/64	05.44	67	19.41	58/60
51.3	Templepatrick	16.35	64	07.35	65	23.14	59
54.4	Doagh	19.31	59	10.19	68	24.55	59
55.9	Kingsbog Junction	21.08	60/58	11.44	65	26.31	60
57.0	Ballyclare Junction	22.07	65	12.36	72	27.28	70
58.3	Mossley	23.08	70	13.39	59 *signals*	28.31	69
59.2	Monkstown	23.58	65	14.35		29.19	69
60.9	Whiteabbey	25.33	69	16.05	78	30.55	73
62.0	Whitehouse	26.27	78 *signals*	16.57	77	32.51	74
65.2	**Belfast**	**30.24**		**21.08**		**35.32**	*signals*
	Net times	29¾ from Ballymena		20¼ from Antrim		32¼ from Ballymena	

Run		**9**		**10**	
Year		6 Sept 1950		20 May 1968	
Train		9.25am ex-Belfast		18.55 special ex-Belfast	
Loco		**56**		**50**	
Load		210 tons		320 tons	
				Piloted to Kingsbog by No 55	
00.0	**Belfast**	**00.00**			
03.3	Whitehouse	06.16	50		
04.3	Whiteabbey	07.40			
05.8	Monkstown	10.43			
07.0	Mossley	13.00	26		
08.2	Ballyclare Junction	15.37	30		
09.2	Kingsbog Junction	17.12		**00.00**	
10.9	Doagh	18.52	63	03.16	56
13.9	Templepatrick	21.35	81		69
15.9	Dunadry	22.54	79/86		75
17.3	Muckamore	23.55		09.13	67
	Post 21	25.06	70	10.18	62/60
19.3	**Antrim**	**26.39**		11.00	64
22.4	Cookstown Junction			14.04	60/68
26.7	Kellswater			18.09	64
31.0	Ballymena			22.18	29/23 *psr*
33.0	Cullybackey			28.00	38 *hand tablet exchange*
38.7	Glarryford			34.21	53/56
40.9	Killagan			37.00	54
43.6	Dunloy			40.06	38 *hand tablet exchange*
	Ballyboyland			45.37	67/68
50.9	Ballymoney			48.16	32/58
54.6	Macfin			53.10	53/58
	Post 59			55.16	50/22/45 *tsr*
57.2	Coleraine			61.23	18 *signals*
62.5	Portstewart			66.28	41/47
65.2	**Portrush**			**72.29**	

Chapter 10
The Visitors

The NCC, though virtually confined to the counties of Antrim and Londonderry, had contacts with other railways which, interestingly, last down to the present day. Since the formation of the RPSI, steam specials on the NCC have at times featured a Great Northern S class and Compound, a DSER mogul and a GSWR J15, to say nothing of the diminutive Londonderry Port & Harbour Commissioners' 0-6-0ST *R H Smyth* which had two lengthy spells of contract work in relaying the lines between Bleach Green and Antrim and Whitehead. This chapter, though, focuses on work done by visiting engines on regular trains in the days of steam.

Great Northern engines

Through excursions from the Great Northern to Portrush via the Antrim branch brought Great Northern 4-4-0s periodically to the NCC. GNRI No 122 is recorded as having piloted Mogul 96 out of Portrush, while Q class No 135 visited Portrush with a Sunday School excursion as late as June 1959 (see photo). On 9 May 1964 no fewer than two S class, Nos 171 *Slieve Gullion* (now preserved) and 174 *Carrantuohill*, came over with specials from the GNRI. No 174 had also worked through from Greenisland to Derry over the 'back line' with an RBAI Railway Society special in 1963 while, in October 1964, U class No 202 *Louth* (as UTA 67) worked another special for that enterprising society from York Road to Antrim, and in September 1965 UG No 149 (as UTA 49) was undoubtedly the only engine of her class ever to penetrate as far as Portrush, on the first RPSI railtour.

Smaller Great Northern engines of several classes, of course, appeared regularly in Antrim, and were used on an early morning commuter service which, until the closure of the Antrim branch to passengers in 1960, worked from Ballymena to Belfast via Lisburn. And, after the UTA takeover, all the Great Northern

In June 1959, ex-GNRI Q class 4-4-0 No 135, joins the NCC main line from the GNR Antrim branch with a Windsor Gospel Hall Sunday School excursion for Portrush.

A Donaldson

Ex-GNRI S class 4-4-0 No 171 *Slieve Gullion* at Portrush turntable on 9 May 1964, on the famous occasion when the redoubtable Fred Graham of Windsor Gospel Hall arranged for two S class engines to work to Portrush.

A Donaldson

engines which received heavy overhaul at York Road works were run in on station pilot duties – some even pressed into service to bank the 8.05pm goods to Ballyclare Junction, to work the perishables train to Larne or to haul local trains from Ballymena to Antrim and Cullybackey.

While the docks lines remained open in Belfast, GNRI RT class 0-6-4 tank No 166 (UTA 24) appeared regularly in York Road yard on trip workings through 'The Subway' from East Bridge Street Junction and along Donegall Quay. This engine, and her three sisters had boiler mountings cut down to enable them to work through the tunnel under the Queen's Bridge.

Until the late 1940s, the NCC's sole experience of broad gauge tank engines was the diminutive J class, so one Great Northern engine that just might have made the NCC think about tank engines was T2 class 'Glover tank' No 142. When virtually new in 1926, this engine was sent to the NCC, and handled for the first week of the trial by a Great Northern crew with an NCC pilotman. Thus crewed, No 142 produced the phenomenal time of 12'45" with a seven-bogie train non-stop to Carrickfergus – a time which no NCC crew could equal, either with 142 or with their own No 83.

Another Glover tank – this time T1 No 187 – appeared on the NCC in 1952, and worked the 5.38pm Belfast–Cookstown Junction local. 'Mac' Arnold recorded the details of this run in *NCC Saga*, but modestly failed to mention that he had been responsible for this surprise choice of motive power. 'Mac' had considerable respect and influence among NCC men, and No 187's main line outing that night was arranged at his suggestion by Inspector Billy Hanley and York Road depot foreman Joe Paul. Preserved S

class No 171 and V class No 85 worked regularly on RPSI specials from the 1970s onwards, but their work is strictly beyond the scope of this book.

County Down engines

BCDR locomotives had a much longer and more regular spell on the NCC during a three year period from 1949 onwards. Inspector Billy Hanley took a personal interest in the County Down engines, and as well as sending Jeeps to the Bangor line, he was instrumental in having some County Down engines tried out on the NCC after the closure of the BCDR main line. (To avoid confusion with the numbers of NCC engines, the County Down visitors are identified by the UTA renumbering scheme, which added 200 to the number of each County Down engine they inherited).

First to arrive was small standard 4-4-2 tank No 212 in April 1949, at the same time as newly-delivered Jeep No 53 was involved in comparative coal consumption tests on the Bangor line. No 212 was used on Larne line locals. In February 1950 she was joined by 0-6-0 No 204 and 4-4-2T No 230 which were steamed over to the NCC. No 230 was put to work on the Larne line, at first shunting at Larne Harbour where she is recorded as having lifted 47 wagons out of the Harbour on a trip goods working to the Town station. No 204 was also used on Larne line trains, and even got as far as Moneymore with a cattle special following the failure of a Scotch engine.

A few weeks later, 4-4-2 tanks 213 and 217 came over from the County Down, followed a month after that by Nos 209 and 221. No 221 was sent straight to Ballymena to join 213 on local workings from that shed, while 209 came to York Road for an overhaul. No 209 emerged from shopping in October 1950, commencing her running-in on Larne line

Above: One of the BCDR's heavy 4-4-2 tanks, No 208, in UTA lined black livery at Whitehead Excursion station in 1950, after working a passenger train. She was running in after overhaul at York Road. *William Robb*

Right: Ex-BCDR 0-6-0 No 4, as UTA No 204, at York Road on 14 May 1950. The UTA had initially given this engine the unusual number '4B', because they already had 2-6-4T No 4 and 4-4-0 No 4A. *HC Casserley, courtesy RM Casserley*

locals, and her place was taken in the works by No 208. After her shopping, 208 worked briefly on the NCC before following 209 back to the Bangor line for what turned out to be the last days of steam on the BCDR.

The engine which most intrigued the NCC was the huge Baltic No 222, the first of a class discussed earlier in this book, and an engine regarded, even by County Down men, as the poorest of the class. How could an engine of such presence have such an absence of power? No 222 was taken into York Road shops for overhaul, and the blastpipe altered to help improve steaming, but all to no avail. The real problem was in the design of the front end. Badly set piston valves and a saturated boiler were not likely to produce a free-running engine. In addition, the system of lubrication was primitive, and could not have coped with fast running. Had there been a future for steam on the County Down, No 222 might have been rebuilt with a G7 superheated boiler and an improved front end. This would, however, have involved expense that could hardly be justified, and after some running-in on the NCC, No 222 shuffled back to the County Down to spend her last years roaring impressively but impotently up and down the Bangor line.

In addition to the passenger engines, 0-6-4T No 229 was a frequent visitor to York Road in the course of shunting around the Belfast harbour area. Indeed, she is credited as having run some 4000 miles on the NCC. She was in the same power class as the NCC 0-6-0s Nos 13, 14 and 15, and was remarkably sure-footed. After 1948 her blastpipe orifice was increased from 3½" to 5" and this was said to have improved her coal consumption, which at times had been in the range of 90lbs per mile.

DNGR engines

The Great Northern engines were visitors, and the County Down engines short-term lodgers, but three ancient little tank engines had a more permanent stay. During the Second World War the NCC was very short of shunting engines, and appealed to the LMS for help. Big brother was too busy with its own locomotives, but someone

in authority remembered that the GNRI had by 1942 largely taken over the working of the Dundalk, Newry and Greenore Railway, and that this small concern now had surplus engines.

So it was that three ancient Ramsbottom 0-6-0STs – tank versions of the famous DX goods engines – arrived on the NCC. No 1 *Macrory* arrived in July 1942, but was in such poor condition that it needed repairs lasting until October before it was fit for use, and saw service at Belfast and Larne. It was finally returned to Dundalk in January 1943, and finished its life being scrapped at Adelaide in 1952. No. 4 *Newry* worked on the NCC from July 1942 until July 1943, and then again from April 1944 until February 1946, mostly in Belfast. No 6 *Holyhead* worked in Coleraine from August 1942 until April 1944. Nos 1 and 4 had 'A' suffixed to their numbers to avoid confusion with the NCC 'Glens', but since NCC 2-4-0 No 6 had been long scrapped, No 6 required no suffix.

NCC men disliked these veterans, most typically for their deficient brake power, and for the screw reverser that was always a nuisance on shunting engines. The three engines which came to the NCC were fitted with vacuum brakes, but these were not very effective. There was one vacuum ejector on the left-hand side of the footplate which created a vacuum on the train but not the engine, and brake power on the engine was controlled by a very powerful handbrake. The regulator was of the one valve type, cylindrical in shape and seated, when closed, in a similar manner to the valves of a motor car engine. The regulator handle, like that of a mogul, could be opened from either side of the footplate. The boiler faceplate had only one gauge glass and a single injector, large and difficult to fit and remove, and operated by pulling a rod which operated a steam cock. It was reliable enough, but if it failed the complete injector had to be sent to Dundalk works as

DNGR 0-6-0ST No 4 *Newry*, one of three locomotives of this type which ran on the NCC during the war.
Kelland Collection, Courtesy Bournemouth Railway Club

149

the barrel was sealed with solder.

In some ways, the DNGR engines were ahead of their time in having self-cleaning smokeboxes which passed soot down the sides of the cylinders next the frames, via a chute to the trackside. They also had an excellent hydraulic gear for starting the pistons from the crossheads for the fitting of new piston rings. In other respects, NCC fitters would have found them less satisfactory. Only a very slim man could get into the firebox, and worse still, out of it again after refitting the firebox door.

Retired Inspector Frank Dunlop must now be the last man to have memories of working on these engines, during the war at Coleraine. Belfast crews would sometimes want to come off, turn their engines and clean their fires over the pit at Coleraine, so No 6 would take the through coaches to Portrush, and work some local services too. Frank remembered that if pressure rose too high, the water in the tanks got warm and the injector became temperamental. No 6 was often driven by Dan Ferguson, an older driver who earlier in the 1930s shared No 94 (the first mogul to come to Coleraine) with Harry Molloy. In his latter years, Dan's eyesight was not so good, and sometimes he was happy enough to let his young fireman have a go at driving the DNGR engine. The mechanical dimensions of these engines are summarised in Table 30.

The 'Jinties'

Since the DNGR engines were not a success, the NCC once again appealed to Derby for help and the response came in the form of 3F 0-6-0 tanks Nos 7456 and 7553, which were renumbered 18 and 19 in the NCC list. They cost the NCC £3,755 each – a bargain bearing in mind that both engines were reboilered in 1944 before delivery. As well as being the NCC's only regauged locomotives, they were each unusual in their way. No 7456 was the only 5'3" gauge Bagnall locomotive ever to have worked in Ireland, and 7553 was the last Hunslet steam locomotive to come to Ireland. The mechanical dimensions are summarised in Table 30.

The engines had slide valves, and despite their relatively long wheelbase (the usual Midland 8'0"+8'6"), the engines could negotiate a four chain curve. Two unusual features for the NCC were the sandbox position – recessed into the tanks for ease of filling – and the provision of smokebox 'dogs' round the circumference of the smokebox (the latter was of course normal on most Deeley/Fowler designs in Britain). Conversion to 5'3" gauge was done simply by reversing the wheels so that they dished outwards and renewing the tyres and crankpins.

Although intended for shunting, the two engines were used on Carrickfergus local trains, until it was discovered that the bearings were inclined to run hot. No 19 is reported to have worked a test train of thirty loaded coal wagons to Ballyclare Junction on three-quarters regulator and the reverser set three notches away from full gear. No 18 worked a similar train, but the lubricator gave trouble on this run.

In April 1950, No 18 was due for overhaul, and the decision was made to shop her in the BCDR works at Queen's Quay. This was one of the last jobs tackled

Far left: Why build an engine when you can have one ready-made? No 18, complete except for wheels, is swung up at Heysham to be shipped to Belfast. Built by WG Bagnall, No 18 is in wartime black but lettered with a proper blocked and shaded 'NCC'.
Author's collection

Left: Now unloaded at Belfast, No 18 has landed on four wheels ready to be hauled to York Road to complete fitting out. Bagnall engines were rare in Ireland. Apart from No 18, they provided only four to the narrow gauge West Clare, though ten more went to contractors and industrial lines.
Author's collection

No 19 doing the job for which she was brought – shunting down the docks, this time on a heavy coal train on rails surrounded by a fine display of square sets and tramway type rail to the left of the engine. Note the shunter's pole across the buffer beam and the man with the flag beside the engine.

Author's collection

No19 at Belfast shed in UTA days on 8 July 1954. She has lost her vacuum brake. No 19 outlasted her sister by seven years. She was built by Hunslet in 1928. Hunslet played a slightly larger role in supplying Irish companies with 19 engines, including the Listowel and Ballybunion Lartique trio. They also supplied 21 industrials.

Author's collection

Left: No 19 on station pilot duties at York Road on 7 October 1961. The tender of mogul No 97 is on the left. The large 'X' on No 19 indicates that she will be withdrawn when the next major fault occurs.

Richard Whitford

Below: Z class 0-6-4T No 27 *Lough Erne* on station pilot duty at York Road about 1968.

CP Friel collection

in these shops. Indeed, in the following month, Scotch engine No 79 *Kenbaan Castle* was also sent round to Queen's Quay. Before work could start, though, the UTA closed the works quite suddenly, and No 79 was towed back to the NCC for completion of her overhaul. No 18 ran 219,441 miles for the NCC (and 612,266 miles in her lifetime) before a suspect crankpin led to her premature withdrawal in 1956. Corresponding mileages for No 19 were 291,971 and 667,521 miles respectively, before final withdrawal in 1963.

The Sligo Tanks

Two of the oddest 'visitors' were the 0-6-4 tanks which the UTA bought in 1959 from Beyer Peacock

following the closure of the Sligo, Leitrim and Northern Counties Railway in 1957. As the firebox was behind the rear driving wheels, these were really a tank version of a Stephenson 'long boiler' 0-6-0, a type originating in the 1840s. Undoubtedly, these engines were the last 'long boiler' engines to be built and also the last to run. These engines renewed the NCC's connection with Beyer Peacock, having been built by that Company in 1949. The financial problems which surrounded their delivery are described in detail in Neil Sprinks' excellent history of the SLNCR. Suffice it to say that a unique form of hire purchase agreement led to each engine carrying on its bunker a plate indicating that it was the property of Beyer Peacock! When the SLNCR closed, the UTA bought both engines, and had them moved by rail from Enniskillen to Omagh (a section of line by then closed to all traffic) and thence to Belfast – not without incident, as one derailed at Fintona Junction. Since the SLNCR never numbered their engines, *Lough Melvin* was numbered 26 and *Lough Erne* 27 and they were designated the Z class. Both engines spent time at Adelaide and then at York Road for shunting around Belfast docks and, apart from three enthusiasts specials, never worked any regular passenger trains. No 26 was withdrawn in 1965 and scrapped in 1968, but No 27 received a full overhaul and the full UTA lined-out black livery in 1965 and passed into the hands of the Railway Preservation Society of Ireland in 1970. Although she was involved in two RPSI specials during her 'company' days, No 27 never ran on the main line in preservation and is currently stored out of use at Whitehead.

A J15 at York Road

For the sake of completeness, it should be noted that the RPSI's preserved J15 class 0-6-0 No 186 was used for a considerable period in 1967 for shunting work in the service of the UTA's successor, Northern Ireland Railways. On at least two occasions in May and August 1967, while based at York Road after overhaul work, No 186 worked NIR ballast trains on the Larne line, and there were up to twenty-five days in the summer and autumn of that year when she was used as York Road pilot, or sent round the Belfast Harbour Commissioners system on dock shunting. On the days when both 186 and No 27 were in use, enthusiasts could see what at that time were the newest and oldest steamable Irish engines, side by side in revenue-earning service! The preserved S class No 171 also shunted at York Road in 1969.

RPSI No 186 at Ballycarry with an NIR ballast train on 11 May 1967. *Craig Robb*

Table 30: Dimensions of non-NCC shunting tank locomotives

Class/numbers		DNGR tanks	Y class	Z class
Type		0-6-0ST	0-6-0T	0-6-4T
Cylinders		17"x24"	18"x26"	18"x24"
Coupled wheels		5' 2½"	4' 7"	4' 8"
Bogie wheels				3'0"
Wheel base		7' 3" + 8' 3"	8' 6" + 8' 0"	6' 7" + 4' 11" +7' 7" + 5' 6"
Boiler length		10' 9"	10' 6"	13' 6"
diameter		3' 9¹³⁄₁₆"	4' 1" (G5½)	3' 10¼"
tubes		??? x 1⅞"	194 x 1¾"	130 x 1⅞"
Heating surface		1069 sq ft	1064½ sq ft	961 sq ft
Firebox		4' 9" x 4'1¹⁄₁₆"	5' 5¹⁵⁄₁₆" x 4' 0½"	5' 7" x 4' 2⅜"
Grate area		15 sq ft	16 sq ft	18 sq ft
Boiler pressure		140 psi	160 psi	160 psi
Tractive effort		13,206 lbs	20,830 lbs	18,890 lbs
Weight		39¼ tons	49½ tons	55 tons, 7 cwt
Water capacity		600 gallons	1200 gallons	1300 gallons
Coal capacity		1½ tons	2¼ tons	2 tons

Chapter 11
The Narrow Gauge

Much has already been written on the NCC narrow gauge, but no work on NCC engines would be complete without reference to the three main three-foot gauge lines and their locomotives. The system comprised three main railways: the line from Ballymoney to Ballycastle; the line from Ballymena to Parkmore and the line from Ballymena to Larne with the branch from Ballyboley Junction to Doagh. In addition there was a three foot gauge tramway from Portstewart station to Portstewart town, the shortest and the first to go, defeated by road competition as early as 1926.

The Independent Companies

We begin with the Portstewart Tram and its three Kitson tram locomotives. Locomotive No 1 was built by Kitson's of Leeds for the opening of the tramway in 1882, to be joined in 1883 by a second, identical, engine, each at a cost of £680. The engines were coke-fired, and designed for one-man operation with duplicate controls at each end. Under Board of Trade regulations for roadside tramways, they were fitted with skirts over the motion and condensers. Following the BNCR takeover of the line in 1897 a third engine, with slightly altered dimensions, was ordered from Kitson in 1900 – almost certainly the last tram engine they ever built. Mechanical details of all three engines are in Table 31. Following the closure of the line, No 3 was sold to David Warke, a Castlerock contractor, who used it until 1935 as a stationary pile-driver. Nos 1 and 2 were taken to Belfast on narrow gauge transporter wagons and remained in store for some years. No 1 eventually passed to the Kingston-upon-Hull Transport Museum in 1939, while No 2 had a more varied retirement. Surviving the blitz on York Road works in 1941, she was later stored in Cookstown Junction shed, then plinthed for a time at Ballymoney, before becoming an exhibit, first in the Belfast Transport Museum, and finally – where she remains today – in the Ulster Folk and Transport Museum at Cultra.

The Ballycastle Railway opened in 1880, and was the last narrow gauge line to come under NCC control when it was rescued from bankruptcy by the LMS(NCC) after a brief suspension of services in 1924. Never well-endowed mechanically, the Ballycastle in independent

Above: Portstewart Tramway No 2, restored to BNCR green livery, being delivered to Ballymoney for display in the late 1940s.

Kenneth Benington

Left: No 113, ex-Ballycastle No 3, at Larne in the full NCC red livery. She has fared better than No 114, getting the blocked and shaded number on the right side of the buffer beam, whereas No 114 had to make do with a plain number. The adhesion factor on these engines was only about 3.2, making slipping very common.

Author's collection

4-4-2T No 114 (ex-BR No 4) is seen here at Larne shed in 1934 with the reduced boiler mountings needed to operate the Ballymena and Larne line. Despite their imposing bulk, these engines were disappointing because of their propensity to slip, first noted on the Ballycastle.

Author's collection

days had to rely on the BNCR for any heavy repairs that were necessary to their engines: and being perpetually short of money, they could only afford the cost of sending particular components to York Road rather than having whole engines overhauled.

A chronological list of the Ballycastle engines appears in Table 33 and dimensions in Table 31. It will be seen that from three small engines with an axle loading of 8½ tons, the Ballycastle Railway suddenly progressed to two comparative giants with axle loadings of 10¾ tons. These 4-4-2 tanks were designed by the Ballycastle's engineer, George Bradshaw, and their major attraction seemed to be a short fixed wheelbase of 6'6" to cope with the severe curves on the line.

Bradshaw submitted the drawings to Bowman Malcolm for inspection, but Malcolm must only have given them a cursory glance, as there were some serious design faults. Only 21 tons 8 cwt of the 39½ tons weight was available for adhesion and this, combined with a high tractive effort of 15,000 lbs, gave an adhesion factor of only 3.2. (The much more successful County Donegal Railways 2-6-4 tanks of class 5A, by contrast, had a tractive effort of 14,300 lbs but an adhesive weight of over 30 tons, giving an adhesion factor of 4.75.) Notoriously prone to slipping, both engines in the words of one exasperated driver "would have slipped on sand". Both engines were fitted with vacuum brakes as delivered – rather an extravagance as the Ballycastle at that time had no braked vehicles, passenger or goods.

For all their design faults, the engines seem to have given the Ballycastle little trouble, and during their sixteen years' work for the independent company

received no heavy overhauls. The LMS was reluctant to scrap such comparatively modern engines, and after trimming them down to the more restrictive loading gauge for the Ballymena and Larne line, put them to work between Larne Harbour and Ballymena. On their adoptive line they were inferior to the NCC's lively 2-4-2 tanks. No 113 returned to the Ballycastle line in 1942, proving no more popular than she had been before her departure.

The Ballymena, Cushendall and Red Bay Railway, a small line with a long name if ever there was one, was built to tap the iron-ore deposits in the North Antrim hills, and ran from Ballymena up to Parkmore. They owned three engines, a chronology of which can be found in Table 33, and dimensions in Table 31. When the Railway was taken over by the BNCR in 1884 they were renumbered, firstly as 60–62, and finally as 101–103 in 1897, when the broad gauge fleet expanded into their number series. In BNCR days No 103 received an extended saddle tank, which she carried until withdrawal in 1911. She was sold to the Antrim Iron Ore Co for its Glenravel Mineral Railway, after overhaul at Larne, but it is unclear when she was scrapped.

Nos 101 and 102 were requisitioned by the government's Irish Railways Executive Committee and hired to the Cavan and Leitrim for the building and early operation of the Arigna extension. On that line they were very popular, though clearly hard-wrought during their nineteen months of service. When returned to the NCC, in November 1921, they were in poor order and, after two years in store as 101A and 102A in reserve stock, they were scrapped in 1924.

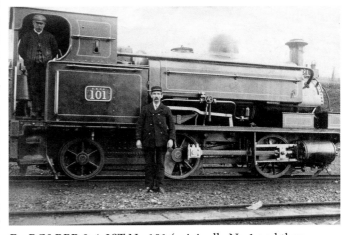

Ex-BC&RBR 0-4-2ST No 101 (originally No 1 and then BNCR No 60 until 1897) has the short saddle tank. Seen here at Parkmore about 1900, the station porter has seized the chance to be photographed with the engine.

Author's collection

Left: 2-4-0T No 64 in 1902, at Ballynure station on its way to Ballyclare and Doagh. *Locomotive publishing Co*

Below left: 0-6-0T No 106, ex-Ballymena and Larne Railway, finished her days at Ballycastle where she is seen about 1930. A serious looking driver stands beside her. *Author's collection*

Below right: No 108 of the same class also made the change to the Ballycastle railway and is seen here at Ballymoney in Sept 1932. *RG Jarvis, Midland Railway Trust, Ltd*

The Ballymena and Larne Railway was the biggest and most progressive of the NCC's narrow gauge lines. Well known for its boat trains, which after 1930 ran with three very modern bogie vehicles, it was the only narrow gauge line to run both boat trains and slip vehicles, about which a few words might be of interest. The 7.55am from Larne Harbour was the express boat train to Ballymena, reached in one hour, and offering a broad gauge connection to Derry. A van of mail for Larne Town was placed at the rear of the train with the chopper coupling connected, but not the vacuum bag. The coupling on the van had a metal ring welded to its front end. A chain was threaded through this ring, and into the window of the van. As the train approached the curve through Larne Town, the driver would make a firm brake application. The unbraked wagon would press against the rest of the train, and a porter in the van would deftly pull the chain to lift the chopper coupling! As the train steamed away up the Inver bank, the mails glided gently into Larne Town. I asked Billy Hanley, the source of this information, whether it was worth all the trouble just to save a minute or two stopping at Larne Town. His reply was that the point of the slip was not to

save station time, but to give the engine the chance of a clear run through the station and on to the Inver bank, which was graded for two miles at 1:36 to 1:44.

This brief digression sets the scene for a study of the Ballymena and Larne engines, chronological details of which can be found in Table 33. The mechanical dimensions appear in Table 32. The 2-4-0 tanks they ordered trace their ancestry back to an engine Beyers produced for the Norwegian Government Railway's 3'6" gauge system in 1866. The Norwegian engineer, R Dhil, must have been satisfied, as eventually a further 26 were built. The precise model for the Ballymena and Larne No 1 was to have been Isle of Man Railway No 1 *Sutherland*, though in the end Beyer's records record that the Irish engine was to be same as No 6 *Peveril*, one of a class still well-known to tourists and enthusiasts, though the B&LR engine was built to a slightly more restricted loading gauge, with the cab three inches lower than *Peveril*'s, and was fitted with the vacuum brake. No 4 appeared a year later and, from experience gained by No 1's first year of service, was modified to carry more water and coal, as well as having more conveniently sited sand boxes.

On the BNCR from 1889, these engines became Nos 63 and 64 and were renumbered 104 and 105 in 1897. In 1889 No 64/105 got 12½" cylinders and a new boiler. No 63/104 was similarly modified with an 88 tube boiler in 1893 and withdrawn for scrapping in 1920. No 105 had a spell of service on the Ballycastle line in 1926, before being sold to the Castlederg and Victoria Bridge Railway in 1928. She was scrapped in 1933.

The 2-4-0 tanks were perfectly satisfactory but, with an adhesion weight of about 15 tons, they were suitable only for passenger trains. The B&LR therefore approached Beyers again in 1877, this time for a more powerful 0-6-0T which would not cause problems on their lightly-laid track. The result was a pair of tanks which had an axle load of only seven tons, but with 21 tons available for adhesion and much more power. Nos 2 and 3 were delivered in 1877 and were followed in 1883 by No 6 which had only minor differences from her two sisters.

After the BNCR takeover in 1889 they were renumbered 66–68, becoming 106–08 in the 1897 renumbering. No 107 was used on the Ballycastle line in 1927–1928, while Nos 106 and 108 seem to have gone to Ballycastle in 1924, spending their remaining days there. All three ran very high mileages for narrow gauge engines: No 106's final mileage was 1,041,899, No 107's was 922,725 and 108's was 857,955.

A great deal of fertile thought must have been going on at Ballymena – we have seen already that the 2-4-0Ts and 0-6-0Ts actually overlapped at the planning stage – and the thinking became even more adventurous between the arrival of No 4 in 1878 and No 6 in 1883. The 2-4-0T was a smooth rider, guided by a pony truck,

whereas the 0-6-0T was more powerful and less prone to slip. Could not the best of both worlds be obtained with a 2-6-0 tank?

Beyer Peacock duly obliged once more, and in 1880 the resulting 2-6-0 saddle tank was a type unique on the Irish narrow gauge and rare anywhere. Although No 5 could be viewed as a stretched 2-4-0T, she was in fact a tank version of a 2-6-0 tender design already in use on the South Australian Railway since 1877.

Despite her bulk, No 5's axle load was less than that of the 0-6-0 tanks. She became BNCR No 68 at first and later No 109. Reboilered in 1899, she acquired the very appropriate nickname 'The Bruiser', and even after Bowman Malcolm's compounds appeared, the men still regarded her as the strongest engine on the line. Their admiration was not, however, unadulterated. Firemen remembered 'The Bruiser' particularly for her cramped cab, where fire cleaning with the rake and the short shovel in a hot and confined space guaranteed a wet shirt and bleeding knuckles!

No 109 accumulated over 900,000 miles, and worked the whole of her career on the Ballymena and Larne section, regularly shedded at Doagh and descending to

Above right: The unusual rear view of 2-6-0ST No 109 at Larne in 1932, showing the brake cylinders and cab.
RG Jarvis, courtesy Midland Railway Trust, Ltd

Right: No 109, the 'Bruiser', is seen here at Larne in May 1924, along with 0-6-0T No 107. She was a tank version of the 2-6-0 tender engines supplied by Beyer Peacock for the South Australian 3'6" gauge.
Ken Nunn Collection, LCGB

Larne shed only when a boiler washout was needed. Her feats of haulage on coal trains from Larne and paper trains between Ballyclare and Ballyboley Junction were legendary, though her great strength was nearly her undoing on one occasion. Billy Hanley was firing to Billy McNinch on a heavy train of thirty loaded cattle wagons. The climb to Ballyboley caused no difficulty, but once over the top the weight of the train took over, and 'The Bruiser' shot down the Inver bank, lurched through Larne Town, and finally stopped on the level stretch to Larne Harbour. Her shaken crew found, on dismounting, that the brake rigging on the engine had broken.

Under the MR(NCC) the inherited narrow gauge engines became classes O–R, in the order of 0-4-2ST, 2-4-0T, 0-6-0T and 2-6-0ST.

The NCC Narrow Gauge Engines

After the absorption of the BC&RBR in 1884 and the B&LR in 1889, the BNCR found itself the owner of nine narrow gauge engines of four different designs. Complete replacement of this collection would have been too expensive, so when new engines were required in 1892, Bowman Malcolm set to work on a standard design of his own. It will be recalled that the BNCR was at this time a committed Beyer Peacock customer, so it was to Gorton that Malcolm looked for a new design of narrow gauge two-cylinder compounds.

There were seven design parameters, including an axle loading of no more than ten tons, a pony truck in front to give smooth running, and coupled wheels of 3'9" diameter, to be standard with the Ballymena & Larne 2-4-0Ts. Oddly enough, given his background and experience, Malcolm asked the builders if they foresaw any difficulties with narrow gauge compounding.

Beyer's draughtsmen had great difficulty with the ten ton axle load limit, and for a time it seemed that the task was impossible. Finally, a drawing was produced which solved the problem. The use of side rather than saddle tanks gave a more even weight distribution, and a reduction in the diameter of the high pressure cylinder, along with a boiler pressed to 160psi rather than 180psi, meant that the desired axle loading was achieved. A chronology of the class appears in Table 33, and dimensions are in Table 32.

The first two were Nos 69 and 70 in 1892, soon becoming Nos 110 and 111. Nos 112 and 113 were built at York Road in 1908–09, but were renumbered 102 and 101 respectively in 1920. The last pair were Nos 103 and 104, built at York Road in 1919–20.

Of the six S class engines, one was rebuilt into class S2, and will be discussed later. Two others became class S1 by the simple addition of a bunker, and four of the class saw out narrow gauge steam on the Ulster Transport Authority in 1950. They were scrapped in 1954.

Some of the class accumulated big mileages. No 111 ran over a million miles in her 58 years of life, despite doing no work at all in the last four years of her nominal life. No 110 covered 614,000 miles as an S class engine, and a further 52,000 as an S2. Smaller mileages were run by Nos 112 (614,000) and 113 (469,761). In their short lives, Nos 103 and 104 ran, respectively, 309,583 and 444,080 miles. When mogul numbers reached 101 in 1939, the 2-4-2Ts were gradually renumbered, Nos 101 and 102 becoming 41 and 42 in that year. No 104 (43) followed in 1942 and finally No 111 (44) in 1948.

The S class engines worked both on their own system and on the Ballycastle line. During her long life

Right: Trains for the Parkmore line were rarely photographed. In the scene which inspired the cover painting for *The Mid-Antrim Narrow Gauge* (Colourpoint 2007), S class 2-4-2T No 111 is leaving Ballymena for Parkmore on 9 August 1930, past a most interesting signal. The train comprises a 1928 bogie coach sand-wiched between two vans.
HC Casserley, courtesy RM Casserley

Left: No 101, with small 'NCC' lettering, at Ballycastle on 29 July 1935 showing the high pressure side of the engine. The extended bunker and longer wheelbase made her Class S1. She too was a Belfast-built engine.

RG Jarvis, Midland Railway Trust, Ltd

Below right: No 102 was the other Class S1 engine. The very large low pressure cylinder can be seen in this view at Ballycastle 1935. Note the traversing jacks carried on the front.

Author's collection

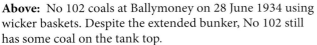

Above: No 102 coals at Ballymoney on 28 June 1934 using wicker baskets. Despite the extended bunker, No 102 still has some coal on the tank top.

W Robb

Right: A feature of the 2-4-2Ts was the carrying of extra coal in large wicker baskets on the apron at the front. No 43 (ex-104 in 1942) is at Ballymoney with her extra coal. She was built in 1920 and was the last Von Borries compound to be built.

Author's collection

Left: No 110 has a substantial train to shift as she passes Larne Harbour signal cabin. In the background the aluminium works can be seen across the road from the engine. Note the mixed gauge crossover in the foreground.

A Donaldson

No 111 was twice reboilered, in 1911 and 1926, and worked on the Ballycastle line. No 42 also saw service on the Ballycastle line, finishing her days working from Larne up to the paper mill at Ballyclare. No 41 finished her life as a popular and efficient performer on the Ballycastle line, while No 43 also went to Ballycastle in 1946, remaining in stock until the line closed in 1950. No 103 was less fortunate, lying out of use for two years after 1936 awaiting an overhaul that never took place. In 1938 it was scrapped.

In 1931, S2 No 110 was radically rebuilt at York Road. The mechanical parts of this engine remained unaltered, except that her trailing truck became a bogie. The resulting wheel arrangement of 2-4-4 tank was unique in Ireland, and highly unusual elsewhere. Bunker capacity was increased to 1½ tons, and the engine became just under four feet longer.

The greatest innovation was a broad gauge G6 boiler, 10'10⅛" long and 4'0" in diameter. To fit into the existing frames, this boiler was shortened to 9'10½" between the tube plates, though it was still longer than the original and 7" greater in diameter.

A new firebox was designed. The outside dimensions were 6'9"x6'2" at the bottom, but to squeeze it into the frames, the inside dimensions had to be no more than 2'5" outside and 1'10⅞" inside. This meant that only a minimal increase of the grate area to 12 sq ft was possible, so that whereas the heating surface of the new boiler was 40% greater than the original, the grate area increased by only 6½%. This could never have been good enough to extract the best performance from such a large boiler.

Grate area aside, the boiler was never likely to be a good steamer. In its broad gauge form, the G6S boiler was superheated with 16 elements and 102 small tubes of 1¾" diameter. As modified for the narrow gauge, the superheater elements were removed and the boiler filled with 164 small tubes – not a recipe for free steaming. The red line of the pressure gauge was optimistically set at 200psi. In the unlikely event of the engine ever having blown off at this figure, the tractive effort would have been 16,000 lbs, but the adhesion weight was only 23.6 tons, giving an adhesion factor of 3.3, which was nearly as bad as the Ballycastle engines.

Above: S class No 110 was rebuilt in 1931 with a saturated broad gauge G6 boiler. She was not a success, being too rigid on the curves. No 110 is seen here at Larne Harbour.

Author's collection

Right: This rear view of No 110 gives an idea of the sheer bulk of the engine, but the adhesion factor was only 3.3, so she was prone to slipping. No 110 was the only 2-4-4 engine in the British Isles and the only NCC narrow gauge engine with a Belpaire boiler, though not superheated. The pressure was later reduced to curb her slipping habits.

HC Casserley, courtesy RM Casserley

Worse again, when pulling hard, No 110 threw herself back on the trailing bogie, and this further reduced the adhesion weight. It was no surprise that this engine derailed frequently, and slipped continuously. Driver John Gamble must have spoken for many when he doubted whether Lough Neagh could have kept her in sand!

In the fourteen years following her disastrous rebuild, the engine ran only 52,000 miles, most of them accumulated whilst shunting at Larne Harbour. She was hardly used at all following the outbreak of World War Two, running only 2,000 miles in 1941, and was not used at all in 1940 or the years 1942–1946. Officially withdrawn in February 1946, she was scrapped a short time later.

In some respects, though, she gave a foretaste of the future. Viewed from certain angles, her Belpaire boiler, cab and Walschaerts valve gear were rather suggestive of the 'Jeeps', which were beginning to appear just as she was leaving the railway scene. It is perhaps appropriate that we conclude our survey of NCC locomotive development with an engine which in appearance at least looked forward to one of the most successful classes ever to run on an Irish railway.

Table 31: Dimensions of narrow gauge tank locomotives

Class/numbers	Portstewart Nos 1& 2	Portstewart No 3	Ballycastle Nos 1 & 2	Ballycastle No 3	Ballycastle T class	BC&RBR O class
Type	0-4-0T	0-4-0T	0-6-0ST	0-6-0ST	4-4-2T	0-4-2ST
Cylinders	8"x 12"	9½"x 12"	13"x 19"	12"x 19"	14½"x 21"	12"x 19"
Coupled wheels	2' 4½"	2' 2¾"	3' 3"	3' 3"	3' 7"	3' 1"
Bogie wheels					2' 6"	2' 7"
Wheel base	5' 0"	4' 6"			5' 0" + 4' 5" + 6' 6"+ 6' 6"	6' 4" + 5' 6"
Boiler length	6' 9"	6' 9"			9' 6"	9' 0"
diameter	2' 5"	2' 7¾"			4' 0"	3' 0¼"
tubes	72 x 1½"		145 x 1⅝"	145 x 1⅝"	170 x 1¾"	104 x 1⅝"
Heating surface	116 sq ft	133.4 sq ft	573 sq ft	573 sq ft	852 sq ft	457 sq ft
Firebox						
Grate area	5.17 sq ft	5.72 sq ft	7 sq ft	7 sq ft	12 sq ft	6½ sq ft
Boiler pressure	150 psi	160 psi	150 psi	150 psi	165 psi	140 psi
Tractive effort	3,436 lbs	5,320 lbs	10,497 lbs	8,945 lbs	13,964 lbs	8,799 lbs
Weight	9 tons	11 tons	24 tons	22 tons	39 tons, 11 cwt	22½ tons
Water capacity			450 gallons	400 gallons	800 gallons	500 gallons
Coal capacity			1½ tons	1½ tons	1¾ tons	8 cwt

Notes: The boiler pressure of Portstewart Tramway Nos 1 & 2 was later reduced to 140 psi and cylinders to 7½" x 12".

Table 32: Dimensions of Ballymena and Larne and NCC narrow gauge tank locomotives

Class/numbers	Class P	Class Q	Class R	Class S	Class S1	Class S2
Type	2-4-0T	0-6-0T	2-6-0ST	2-4-2T	2-4-2T	2-4-4T
Cylinders	11"x 18"	13½"x 18"	14"x 18"	14¾/21"x 20"	14¾/21"x 20"	14¾/21"x 20"
Driving wheels	3' 9"	3' 3"	3' 3"	3' 9"	3' 9"	3' 9"
LW/TW	2' 0"		2' 0"	2' 0"	2' 0"	2' 0"
Wheel base	8' 0" + 6' 3"	5' 0" + 5' 6"	5'6"+3'6"+3'6"	6'9"+6'3"+7'3"	6'9"+6'3"+9'3"	6'9"+6'3"+8'3"+4'6"
Boiler length	7' 8¼"	8' 0"/8' 4½"	9' 2"	9' 4"	9' 4"	9' 11" (G6S)
diameter	2' 10¾"	3' 3"	3' 5¾	3' 5¼"	3' 5¼"	4' 0"
tubes	103 x 1⅝"	132 x 1⅝"	160 x 1⅝"	150 x 1⅝"	150 x 1⅝"	164 x 1¾"
Heating surface	392 sq ft	507/529 sq ft	695 sq ft	673 sq ft	673 sq ft	825 sq ft
Firebox						
Grate area	6.95 sq ft	9.1 sq ft	10½ sq ft	11.3 sq ft	11.3 sq ft	12 sq ft
Boiler pressure	140 psi	140 psi	140 psi	160 psi	160 psi	200 psi
Tractive effort	5,759 lbs	10,009 lbs	10,765 lbs	13,150 lbs	13,150 lbs	16,438 lbs
Weight	17tons, 12ct 19tons, 3¾ct	21tons, 2ct 22tons, 9½ct	25tons, 13½ct	31 tons, 17 cwt	33 tons	42 tons, 1 cwt
Water capacity	385/450 gallons	450 gallons	500 gallons	570 gallons	570 gallons	570 gallons
Coal capacity	1/1.3 ton(s)	1/1.1 ton(s)	1.2 tons	1 ton	1 ton	1½ tons

Notes: The coal bunker on Class S1 had nominally the same coal capacity as the S class, but kept the cab clear of coal.

Table 33: Chronology of the narrow gauge locomotives

Class	Type	No	Deliv	Builder	Cost	1897 No	Rebuilt	Renumbered	Scrapped
O	0-4-2ST	60	4/1874	BH 301	£1432	101	1897, 1907	101A (2/1920)	?/1923
O	0-4-2ST	61	4/1874	BH 302	£1470	102	7/1893, 1908	102A (2/1920)	?/1923
O	0-4-2ST	62	4/1875	BH 303	£1470	103	3/1893		2/1911
P	2-4-0T	63	3/1877	BP 1687	£1432	104	1893		2/1920
P	2-4-0T	64	10/1878	BP 1828	£1432	105	1889		sold 1928
Q	0-6-0T	65	8/1877	BP 1700	£1560	106	1897, 1908		6/1933
Q	0-6-0T	66	8/1877	BP 1701	£1560	107	1898, 1912		12/1931
Q	0-6-0T	67	2/1883	BP 2304	£1610	108	1898, 1906		12/1932
R	2-6-0ST	68	5/1880	BP 1947	£1700	109	1899		5/1934
S	2-4-2T	69	5/1892	BP 3463		110	2/1910 9/1931 (S2 2-4-4T)		2/1946
S	2-4-2T	70	5/1892	BP 3464		111	1911, 1926	44 (12/1948)	2/1954
S	2-4-2T	112	10/1908	York Road		–	6/1930 (S1)	102 (2/1920) 42 (11/1939)	2/1954
S	2-4-2T	113	3/1909	York Road		–	7/1928 (S1)	101 (2/1920) 41 (6/1939)	2/1954
S	2-4-2T	103	9/1919	York Road		–			12/1938
S	2-4-2T	104	3/1920	York Road		–		43 (10/1942)	2/1954
T	4-4-2T	113	9/1908	Kitson 4565		–	12/1926		7/1946
T	4-4-2T	114	9/1908	Kitson 4566		–	11/1927		2/1942

Notes: T class No 113 was withdrawn in 9/1940 and reinstated in 2/1942.

Chapter 12
Diesel Traction on the NCC:
Locomotives and Railcars

Diesel traction played a very minor part in NCC locomotive history and the story of diesel development in Ireland in general, and the NCC in particular, is beyond the scope of this volume. Irish diesel locomotive development has been covered with customary thoroughness by Michael Rutherford in 'Emerald Isle Innovation: dieselisation in Ireland' (*Backtrack*, November & December 2001). Students of diesel railcar development in Ireland could learn much from Colm Flanagan's *Diesel Dawn* (Colourpoint, 2003) and Michael Collins' *Road versus Rail in Ireland 1900–2000* (Colourpoint 2000). This chapter offers only a very brief summary of the salient points.

Locomotives

Until the mass dieselisation of CIE in the 1950s, diesel locomotives (apart from the many narrow gauge industrial units used by Bord na Mona in their extensive peat bog systems) were almost unknown on Irish railways; indeed only one Irish firm – the world famous shipbuilders Harland & Wolff of Belfast – ever built any. Harland's Diesel Department was set up in the 1930s when their main business was hard-hit by world depression, and the company built locomotives that saw service in Argentina, Canada, New South Wales and The Sudan.

Harland & Wolff's main Irish customers were the BCDR and the LMS(NCC), and they delivered two small diesel locomotives to the BCDR in 1933 and 1937, both intended to work passenger trains. The first was a 250bhp diesel mechanical shunter numbered D1, whose history is of no relevance to this chapter. However, the second was a double-bogied 500bhp diesel electric locomotive, initially numbered D2, but almost immediately renumbered 28. It is worth pointing out that, as far as is known, No 28 was the first diesel locomotive in the British Isles to run on bogies.

As the BCDR hired, but never actually purchased this engine, it went back to Harland & Wolff in December 1944, and was hired on to the NCC in 1945, working on both the main line and the Larne line until 1952. Apparently well thought of, it ran on light passenger trains at speeds of up to 60mph, though it ultimately became a purely shunting locomotive with its speed strictly governed.

No 28 had no train-heating equipment (nor in fact did any Irish diesel locomotive), and the NCC anticipated later CIE practice by building a 20 ton steam heating van to run with it. Around 1952 the locomotive was returned to Harlands, but in 1957 was hired once again to a third company – this time the GNRB. When the GNRB was dissolved the following year, the UTA took over the hiring agreement. The UTA finally bought No 28 in 1962, and it spent the rest

Diesel locomotive No 28 arriving at York Road with a Larne line local around 1951. This locomotive had the unique distinction of carrying the same running number successively on the BCDR, NCC, UTA, GNRB, UTA (again) and NIR. Built in 1937, it was one of the first diesel locomotives to have power bogies rather than a rigid 0-4-0 or 0-6-0 wheel arrangement.
Lens of Sutton

of its life shunting at Great Victoria Street station and Grosvenor Street goods yard. When scrapped in 1973, now in NIR ownership, No 28 had the distinction of having worked for five different companies without a change of number!

Although No 28 was a well-known part of the Belfast railway scene, the NCC's other diesel locomotives were much less obvious, and indeed few photographs are extant of what Colm Flanagan appropriately dubbed the 'shadowy shunters'.

The first to be built originated as a 150bhp diesel-hydraulic 0-6-0 shunter which Harland and Wolff built for the LMS in 1934. Harlands were one of a number of builders who supplied diesel locomotives to the LMS that year for appraisal. As LMS No 7057, it worked at Chester and Heysham, before being sent back to Belfast in 1944, probably due to its non-standard nature and the difficulty of obtaining spare parts locally in wartime. Harlands re-gauged it, and re-engined it with a 225bhp engine and, as NCC No 22, it spent the rest of its days creeping round York Road, governed to a top speed of 10mph. Having the possibly unique distinction of being the only Irish-built diesel to have

run on the 4'8½" gauge in Britain, it was scrapped by the UTA in 1965.

A more powerful locomotive was No 17, another Harland & Wolff product, which the NCC hired in March 1937 and bought in 1941, designating it 'class X'. A 0-6-0 diesel mechanical, it was of 330bhp and could deliver a starting tractive effort of 24,000lbs. No 17 was a powerful and popular shunting unit, which at one time was shunting Belfast yard continuously, apart from a six hour break each Monday for maintenance. Although No 17 technically passed into the hands of NIR in 1967, it saw little use, if any, after March 1966 and was scrapped around 1970.

The final unit of this trio was the diminutive No 16, a diesel mechanical 0-4-0 of 225bhp, apparently with the same power plant as the one later fitted to No 22. Its curious appearance derived from the fact that it was built by Harlands in 1937 as a works shunter in the shipyard, where a four-coupled engine with a long wheelbase was essential to cope with the sharp curves into some of the engineering shops. It was first used by the NCC in 1945–1946, temporarily numbered 20, but was returned to its makers when No 22 arrived.

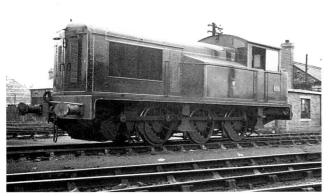

Upper left: No 17 in fresh UTA livery in May 1950. Note the prominent jack shaft.
Kelland Collection

Upper right: No 16 in typical grimy condition at York Road. This engine also had a jack shaft, unusual in a 0-4-0.
CP Friel Collection

Left: No 22 pauses for the benefit of the photographer at York Road in September 1946. As LMS No 7057 it had shunted at Chester and Heysham from 1934 to 1945 and was a rare example of a locomotive being built in Ireland for service in England.

AW Craughton

In 1951 it came back into railway hands, either sold or exchanged for the ex-BCDR 'D1' (BCDR No 2 since 1937) which was of no further use to the UTA, following the closure of most of the BCDR. On rejoining the NCC fleet, it was renumbered 16, taking the number of 'The Donkey', the celebrated York Road 0-4-0ST shunter which was described on page 74. No 16 was finally disposed of in 1967, though like its two sisters it saw very little use in its latter years.

Railcars

Leaving aside the NCC's two railbuses (which were converted road vehicles rather than units built specifically for railway service), we come now to the railcars built from 1933 onwards.

In some ways, the railcars were the successors to the steam railmotors described earlier, and they were certainly more commercially successful. Of the Irish standard gauge companies, the GNRI and the NCC made most use of railcars in the 1930s, and the NCC's first example was railcar No 1.

Built at Belfast in 1933 as a joint venture with the Leyland Motor Company, railcar No 1 had underfloor engines driving the axles of end-mounted bogies, and was powered by two 130bhp Leyland petrol engines (virtually identical to engines which that company developed for road buses). Roof-mounted radiators were fitted – a practice common on the early railcars but not generally perpetuated after World War Two. Railcar 1 was re-engined in 1947, and again in 1959 with diesel engines similar to those used on the MED railcars.

It was an extremely successful unit and, as it was mounted on coach bogies, it was comfortable to travel in. Though designed with the less-trafficked branches in view, it spent much of its working life on suburban workings from Belfast, and on the Portrush branch. Two lightweight railcar trailers (weighing only 17½ tons, though theoretically seating 100 passengers) were built in 1934–35 to

Above right: No 2 in original condition at York Road in 1934. Its first livery was plain LMS red. This was the first diesel railcar on the NCC.

LGRP

Right: No 2 at Portstewart in red and white livery after its original driving turrets were removed.

Real Photographs Ltd

Railcar No 1 at York Road in 1936. Although petrol-engined until 1947, this railcar had the twin-engined under-floor layout that was to become standard on all post-war railcars in the British Isles.

Real Photographs Ltd

run with this and the later railcars.

One of railcar 1's last duties was driver route-learning on the ex-GNR Lisburn–Antrim branch in 1964. The closure of the GNRI 'Derry Road' the following year was going to lead to the re-routing of Dublin–Derry freight traffic, and NCC drivers needed to be passed out over the Antrim branch. After 1965, No 1 was stored serviceable at Ballymena, and never worked regularly again. Destined for preservation, it passed to the custody of the RPSI, but must be regarded as a very long term project, in terms of restoration.

In 1934 railcar No 2 appeared, this time with twin Leyland diesel engines. To balance the increased weight

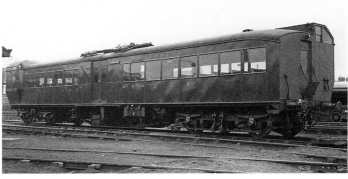

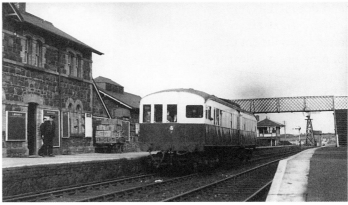

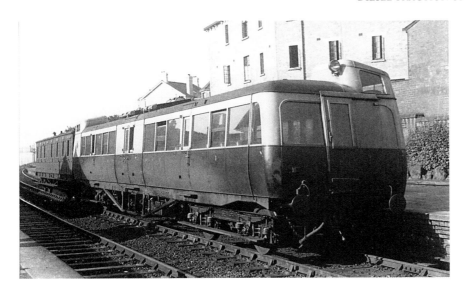

Left: Railcar No 3 at Whitehead with a local from Belfast in NCC days, sporting the LMS red and white livery. It seated 12 first and 60 third class but this has been supplemented by an old bogie carriage. The small turret was the driving position.

Real Photographs Ltd

Below: The Ganz railcar at Queens Quay station, Belfast, on 23 April 1951, its first day in service. The engine was above floor, behind the driver. Note the McCreary tram in the background. Both Queens Quay and York Road had tram terminals inside the station.

TJ Edgington

of the engines, the bodywork of the car was six tons lighter than that of No 1, despite being some six feet longer. So light was the body, in fact, that it buckled whilst being lifted for fitting on to its chassis. Raised turret cabs and end doors were provided at each end, as the original idea was that the car could pull or propel a trailer (or even run sandwiched between two). The raised cabs were eventually removed but No 2's extremely lightweight construction was probably responsible for its early withdrawal in 1954. The bogies, engines and transmission were made into two tractor units used to lift the ex-BCDR main line, while the body survived until the mid 1960s as a classroom for fire training at York Road. Ironically, it was eventually gutted by an accidental fire!

Railcars 3 and 4 were virtually identical vehicles, No 3 having rather superfluous (and very narrow) end doors. Both had raised turrets at each end which provided extremely narrow cabs for drivers, though after a derailment on the Portrush branch the practice of propelling railcar trailers was banned. Railcar 3 was destroyed by an accidental fire at Whitehead in July 1957. Railcar 4 became the last of the NCC railcars to see regular use, being last used in 1966 though technically surviving into NIR days. It suffered the same fate as its sister, this time in a fire at York Road in 1969.

The last railcar to be mentioned had, strictly, only an indirect connection with the NCC, as it was purchased

by the UTA in 1951 as railcar No 5. As far back as 1937 a single unit diesel railcar was built by Metropolitan Vickers and the Metropolitan Cammell Carriage and Wagon Company, under licence from the Hungarian firm of Ganz. The car was similar to the Arpad vehicle used on a fast service between Budapest and Vienna, but although it could work at speeds of up to 70mph, its seating capacity was restricted (to only 36 third and 18 first class passengers) by the very large engine, encased in a polished wood cover, which occupied part of the passenger saloon behind the driver's cab. Known as the 'Metro Vick Cammell Car' it was initially tested between Euston and Tring on the LMS, and in July 1937 *The Railway Gazette* said that "arrangements are now being made for it to go into regular operation in this country."

Its development seemed to be linked to an anticipated boom in the overseas market for railcars, particularly in South America – indeed the year before

Left: Railcar No 4, in its final condition, at York Road in April 1964. It continued to work locals to Ballymena until early 1966. It was then stored awaiting a decision as to its future but was destroyed in a malicious fire in 1969.

Roger Holmes,
*Courtesy **Photos of the Fifties***

Below: In UTA days, railcar No 1 is seen approaching Whiteabbey with the 2.40pm local service from Ballymena on 14 September 1963. This pioneer railcar has been preserved by the RPSI.
Derek Young

its appearance, a block order for 100 Drewry railcars had been placed by Argentina. The 'Ganz' created no such crock of gold for any British manufacturer, and it had a rather peripatetic existence. Further tested in the north-east, where it was exhibited at Hull and elsewhere, the Ganz car spent the war in seclusion, probably at Metro Cammell's Birmingham works. Thereafter it was regauged and sold off to the UTA in 1951.

By this time, the UTA was thinking in terms of dieselising the Bangor line and the NCC and, on delivery, No 5 featured in comparative trials involving one of the NCC railcars, and Harland & Wolff diesel locomotive No 202 (ex-BCDR No 2). Although no records have survived, the NCC railcar seems to have made the best impression. The Ganz car worked on the Bangor branch as well as the NCC, and was transferred to the Great Northern section after that company's dissolution in 1958. An eccentric 'one off' was of no great use to an area intensively worked by fairly modern AEC and BUT multiple units, so No 5 spent most of its latter years in store at Adelaide, before scrapping in 1965. Its dedicated trailer, built by the UTA in 1954, survived as an MED intermediate vehicle until 1980.

In NCC service, railcars 1–4 at first carried maroon coach livery, later maroon with cream upper panels. In UTA service, they were painted in that company's rather drab green, with cream upper work. Unlike the railcars which the UTA later inherited from the GNRI, the NCC cars were never renumbered, and from the early 1950s the UTA's own build of new railcars followed in sequence with No 6 in a series which was kept separate from other coaching stock. The Ganz railcar carried UTA green and cream livery throughout its life in Ireland.

Table 34: Dimensions of NCC diesel locomotives

Class/numbers	Class X No 17	No 20/16	No 28	No 22
Type	0-6-0	0-4-0	1A-A1	0-6-0
HP	330hp	225hp	500hp	225hp
Wheels	4' 0⅛"	3' 7"	3' 7"	3' 2"
Wheelbase	6' 0" + 6' 0"	8' 0"	7' 0" + 8' 6" + 7' 0"	6' 0" + 6' 0"
Cylinders	135x220mm	135x220mm	180x300mm	135x220mm
No of cylinders	Eight	Six	Eight	Six
Transmission	Mechanical	Hydraulic	Electric	Mechanical?
Weight	49 tons	28 tons, 6 cwt	48¾ tons	27 tons, 3½ cwt

Notes: As built, No 22 had a 175hp four cylinder engine (148x220mm) and weighed 27½ tons.

Table 35: Chronology of Harland and Wolff diesel locomotives

No	Type	Built	Acquired	Comments	Withdrawn
17	0-6-0	3/1937	3/1937	Hired 1937–41, purchased 1941	?/1968?
16	0-4-0	1944	1/1945	H&W Works shunter. Hired 1/1945–4/46 as No 20. Bought by UTA 1951	?/1965?
28	1A-A1	1937	7/1945	Hired 1945–52	9/1952
22	0-6-0	7/1934	1/1946	LMS No 7057 from 2/1935 to 1/1944. Returned to H&W 1/1945. Regauged 1945. Hired to NCC 1/1946. Bought by UTA 10/1949.	4/1965

Notes: As LMS No 7057, No 22 operated at Chester and Heysham.

When built in 1937, No 28 was the first bogie diesel-electric locomotive in the British Isles.

Table 36: Chronology of NCC railcars

No	Type	Built	Comments	Withdrawn
1	Petrol	1/1937	Re-engined 1946?. Preserved at Whitehead	?/1968
2	Diesel	6/1934	Body converted to a lecture room at York Road.	?/1954?
3	Diesel	1935	Withdrawn due to an engine fire at Whitehead.	?/1956
4	Diesel	7/1936	Last used 1966	12/1969
5	Diesel	1939	The 'Ganz'. Bought by UTA in 1950	5/1965
42	Railbus	12/1934	Reg No CH 7910. Not used after 1939.	1942
43	Railbus	5/1936	Reg No CH 7913. Not used after 1939.	1942

Notes: Nos 42 and 43 carried road fleet numbers.

Chapter 13
NCC Steam in Preservation

Enough has been said to indicate that the NCC was, until the late 1940s, a railway of considerable locomotive variety. By 1947, though, the 'Glens' had all gone, as had the 'Whippets'. Some A1 class engines survived until 1954, latterly seeing little use. The coming of the mogul tanks, along with the closure of the Derry Central line and most of the branches in 1950, displaced the older engines, and the slaughter of the 'Scotch' engines began in 1956; half the class had gone by 1957 and the rest by 1961.

Sadly, the NCC did not fare well in preservation and, of an estimated 36 Irish locomotives still in existence around the world today, only four NCC engines still exist – Nos 4 and 74, and Portstewart Tram engines Nos 1 and 2. It might be remarked, in passing, that New Zealand, a country of comparable size to Ireland, has over 80 active preserved engines, with another dozen awaiting overhaul. For the record, seven NCC coaches, one railcar, six wagons and one hand-crane also survive in preservation.

None of the NCC 4-4-0s might have been preserved, had it not been for the efforts of Harold Houston, who ensured that U2 class No 74 *Dunluce Castle* was saved for preservation, cosmetically restored in the full glory of NCC lined maroon livery. Typical of his attention to detail was the fitting of a false capuchon to disguise her historically incorrect Stanier chimney. Displayed for some years in the original Belfast Transport Museum at Witham Street, No 74 (along with all the other exhibits) was delivered to the railway gallery of the Ulster Folk and Transport Museum on Sunday 14 February 1993. This gave No 74 her first main line outing in over three decades, as she and GSR 4-6-0 No 800 *Maedhbh* were hauled dead – appropriately by preserved GNR 4-4-0 No 171 – from Adelaide yard to Cultra.

Sadly, not one mogul was preserved, though there was some speculation about No 97 in 1965. After she was scrapped, her boiler lay at York Road until the end of steam. Of the Jeeps, Nos 4 and 53 were stored at Carrickfergus after the end of steam on NIR in May 1970 and, had the money been available, both might have been acquired by the RPSI. No 53, indeed, was displayed out of steam at Whitehead at an RPSI Open Day on 27 June 1970, but it was No 4 which was moved to Whitehead permanently, officially becoming the RPSI's property on 11 July 1971 at a cost of £1275.

No 4 has now had a longer life in preservation than in company service, and her wheel arrangement makes her unique among Irish broad gauge preserved locomotives. In the course of 37 years of traversing the Irish railway system she has hauled trains to destinations as varied as Cork, Limerick, Tralee, Galway, Sligo, Westport, Waterford and Rosslare, setting new standards of performance wherever she has gone. Older CIE men pronounced her as good as anything they ever stood on; younger men said she was better than the A class diesels that saw off steam on CIE in the 1950s!

In preservation No 4 currently has blank tank sides, but in their lifetime the jeeps carried no less than four company designations: the rather ugly block-lettered 'NCC', followed by the UTA 'red hand' emblem and after 1959 an armorial device granted by the College of Arms. Its motto 'Transportatio Cultum Significat' (Transport signifies culture) might have given pause for thought to passengers in the mid 1960s, seated in MED and MPD railcars which had suffered failures! The final emblem carried by a few of the jeeps was that of Northern Ireland Railways, a company which in the year of publication of this book celebrates its fortieth anniversary.

Steam finished in Northern Ireland in 1970, but between June and November 2000 the very welcome decision to relay and re-open the derelict NCC main line between Bleach Green and Antrim brought steam back on a daily basis. Henry Boot, the main contractors, hired ex Londonderry Port & Harbour Commissioners 0-6-0 tank No 3 *RH Smith* from the RPSI. In the course of 90 days of working, she shifted some 50,000 tons of ballast without a single failure, and played a distinguished part in restoring this part of the NCC to 90mph running standards. Between August and December 2005, No 3 had another 90 days of hard NCC work, this time on the AMEC relay of the Larne line between Bleach Green and Whitehead. Since these were absolute possessions under the control of contractors, the engine was crewed by qualified volunteers from the RPSI and Downpatrick.

The last word, appropriately, is with steam. During the hardest days of World War Two, only six total locomotive failures occurred on the NCC between 1940 and 1945. Could any modern railway make a similar boast?

Appendix 1
Liveries

BNCR

The very earliest photo of a BNCR locomotive is the one of G class No 40 on page 31, showing the number painted on the cab sidesheet and the letters 'BNCR' on the tender, but centred on the rear panel. This probably reflects the livery up to the end of the 1870s, in the era when there were no cabs, only weatherboards. This reflects B&BR practice as suggested by Harold Houston's drawings on pages 12–13.

By the early 1880s the engines were painted a very dark green, lined crimson, blue and yellow, with the letters 'B N C R' on the tender, well spaced out. The engines now all carried numberplates. The only other change in BNCR days was that, from 1897 on, passenger engines carried the company crest on the cabside.

MR(NCC)

When the Midland Railway took over, the engines were painted 'invisible green' – almost black – once again lined in crimson, blue and yellow. Black smokeboxes and red buffer beams completed the engines. In early Midland days, BNCR practice was continued by having the Midland crest on the cabside and the letters 'N C C' spaced out on the tender. A change was made about 1904–05 in that the crest was moved to the leading splasher.

After the First World War the livery changed. The crest was back on the cab, and the tender now bore the letters 'M R' in gold capitals, with 'NCC' in scroll between the M and R. With the new locomotives built in 1922–23 there was a return to pre-war style with 'N C C' on the tender and the crest on the splasher.

LMS(NCC)

The LMS introduced the crimson lake (maroon) with the LMS crest on the cab and the letters 'N C C' on the tender sides. Over the next twenty years, lettering style followed the fashions and fonts used on the parent company. For example, the Derby built moguls in 1933 carried very large 'N C C' letters on the tender, but from 1934 a smaller size prevailed.

Up to the end of 1927, it was the practice for all LMS passenger engines, including tanks, to be painted in the full red livery. From 1928, for reasons of economy, it was decreed that only *Express* engines like the 'Royal Scots', Pacifics and 'Jubilees' should have the full livery. The NCC, however, continued to paint all passenger engines in red, except some older ones that retained 'invisible green'. This might have been justified for the 'Glens' and 'Castles', since they were 'principal passenger engines', but hardly for the 'Counties' and 'Mountains', still less the C1 class 2-4-0 compounds!

The Moguls, which were as much mixed traffic engines as the LMS 'Black Fives', also got crimson lake on the basis that they were now the 'principal passenger engines', but the 'Castles' still stayed in red! Even the war time moguls appeared in red.

It was not until the closing stages of the Second World War that NCC engines were painted black, with the exceptions mentioned in the text. After 1945 a lined black livery was introduced, reflecting post-war LMS practice. Lining was a single red line, with the LMS crest high up on the cab, as seen on No 74 on the rear cover. However, the WT class 2-6-4Ts had a different lining using straw and maroon, as described on page 125.

UTA

Black became the standard livery of the UTA, but it was a much brighter black than the NCC version, more like the 'Blackberry black' of the old LNWR. UTA lining was much more elaborate with parallel straw and red lines. To add to the colour, the flutes of the coupling rods were also painted red, as were the flutes of the motion and connecting rods of moguls and Jeeps. However, the coupling rods of these two classes were not fluted and so not red. All this looked superb when engines were ex-works. The first few repaints appeared with a large 'U T' on the tender but, from early 1950, the UTA 'red hand' roundel appeared on the tender or side tank, rather than any lettering.

In 1959 the UTA roundel was superseded by an armorial device which depicted a shield supported by a red lion and brown stag, with a medieval helmet, oak leaves and winged horse for good measure. Below was the Latin inscription "Transportatio Cultum Significat", roughly translating as "Transport signifies culture". In other respects, the livery was unchanged.

NIR

The final contribution to liveries was when NIR in 1969 replaced the heraldic device with a gold-leaf NIR symbol on the side tanks of a few of the Jeeps.

Number plates

BNCR locomotives do not appear to have carried numberplates until the 1880s. The BNCR, and later the NCC, then used rectangular brass number plates with radiused corners, and ornate figures. These were given even to the diesel shunters! The major peculiarity of number plates was the casting of the number '7' which resembled an inverted '2'. The 'Jeeps' carried much plainer numberplates without any embellishments.

Builders' plates

Makers plates varied from the famous and handsome 'Beyer Peacock' wording round the splashers of the older engines, to the less obvious builders plates which appeared in various places depending on the class of engine. Engines built at York Road had an oval plate with the company name at the top, then 'Belfast' followed by the date. Those built in England, of course, carried the name 'Derby' rather than Belfast.

Appendix 2
Engine Headlamps

The BNCR had a complex system of lamp codes. By day, branch trains carried a white diamond at the chimney and by night, or in foggy weather, trains carried a combination of red, white and green lamps which allowed signalmen to differentiate goods trains, main line passenger trains and the destinations of branch trains at Junctions.

However, as the following extract from the 1931 Working Timetable shows, the NCC still continued to use codes that were different from LMS practice.

The NCC superseded this with the standard English system, reproduced opposite. In 1953, mogul No 102 had to have an extra lamp bracket fitted so that she could carry the regulation three lamps across the buffer beam for the Royal Train.

In theory (possibly a wartime expedient) this code was replaced by a simpler one in 1943, whereby all train engines were supposed to carry only one lamp at the chimney and light engines, and engines running round, also carried a tail lamp. In practice, though, almost every photograph taken in the 1940s, 1950s and 1960s shows a 'proper' lamp code. Passenger trains carried a tail lamp on the last coach, with a second lamp above it to indicate if a special train was following.

ENGINE HEAD LIGHTS. 1931

For the guidance of Signalmen and others, the undermentioned Head Lights must be carried by Engines working Trains or running light after sunset and in foggy weather :—

Main Line Trains on Double or Single Line	.. White Light on Chimney.
Trains for Branch Line on Double or Single Line	.. White Light on Chimney and White Light over right-hand Buffer in direction of running.
Derry Central Branch Trains	.. White Light on Chimney and White Light over right-hand Buffer in direction of running.
Draperstown Branch Trains	.. White Light on Chimney and White Light over left-hand Buffer in direction of running.
Narrow Gauge Trains, including Parkmore Branch ..	Purple Light on Chimney.
Narrow Gauge Trains for Branch, on Main Line	.. Purple Light on Chimney and White Light over right-hand buffer in direction of running.

Trains running from a Main Line to a Branch Line without stopping at the Junction Station will change Head Lights at first stopping Station after entering on Branch Line.

Branch Trains will be distinguished by a White Diamond-shaped Board at foot of engine chimney by day.

TRAIN ENGINES STANDING BEYOND PLATFORM STARTING SIGNALS AT BELFAST.

When engines of trains are standing beyond the platform starting signals at Belfast, the Station Master, Inspector or Station Foreman must personally inform the Driver when the Signal has been turned off for the train to proceed.

ENGINE HEAD LIGHTS.

In order that Signalmen and others concerned may be able to distinguish the class of Train that is approaching by the Head Lamps on the Engine, all Engines running over this Committee's Line must, unless instructions to the contrary are issued, carry white head lights arranged as under. The Lamps must be carried in the authorised position during daylight.

	Head Lights.
1. Express Passenger Train, Breakdown Van Train going to clear the line, or Light Engine going to assist disabled Train	
2. Ordinary Passenger Train, Rail Car, or Breakdown Van Train not going to clear the line	
3. Fish, Meat, Fruit, Milk, Horse, Cattle, Perishable Trains composed of Coaching Stock or Empty Coaching Stock Trains	
4. Through Freight, Mineral, Cattle or Ballast Train ..	
5. Freight, Mineral, Cattle or Ballast Train stopping at intermediate stations	
6. Light Engine or Light Engines coupled together, or Engine and Brake	
7. Shunting Engine—Red Head and Tail Lights ..	

Appendix 3
Gradients and Enginemen on the NCC

The Broad Gauge Lines

In the down direction, the principal gradient on the NCC was the climb from Whitehouse to Kingsbog Junction. Ten bogies was the normal unpiloted load, but on occasion eleven was taken, sometimes without the driver's knowledge.

Towards the end of steam, for example, the late RJ ('Batman') Simpson had No 53 on a special to Portrush. The train was eleven bogies, and no other engine was available to pilot the train. So Inspector Frank Dunlop 'sang dumb' about the load, innocently suggesting that there must be a few brown vans against the buffers when the driver remarked that the front of the train seemed to be well off the end of the platform. The first significant curve on the line does not occur until Bleach Green Viaduct, and here drivers always looked back to check their load. Those of us in the train first saw Simpson leaning well out of the cab with an expression of disbelief, and then his face turning scarlet with rage as he suddenly realised why his engine seemed to be pulling so sluggishly! He was a great engineman, though, and once his legendary mercurial temper subsided, he saw the funny side of it, and produced an extremely good run given the load and the condition of the engine.

Another story deserves to be recounted, this time of preserved ex-GNR three-cylinder compound No 85. Diplomatically described by driver Tom Crymble as "an interesting engine", this sometimes brilliant though often temperamental machine was the cause of much head-shaking among the small group of ex-NCC drivers still in service when No 85 was restored to working order in the mid 1980s.

In the course of one eight-coach running-in trip from York Road to Queen's Quay via the Antrim branch, No 85 was plodding steadily up the bank tender-first with the experienced crew of Harry Ramsey and Davie McDonald. When they were down to walking pace above Monkstown, Inspector Frank Dunlop called across the cab: "Give her another turn, or she'll never get up", only to receive the retort, "There is no more turns, Frank!" In fairness, No 85 did make it, though the low-pressure cylinder steam-chest gauge was hovering around 50lbs when she was working in compound. The other man on the footplate that day, though, rightly claimed to "know about compounds". Davie McDonald started his days in Larne, where he

had fired the small narrow gauge compounds up the Inver.

Once past the top of the hill, pilot engines of heavy trains were removed either at Ballyclare Junction or Kingsbog Junction, the latter having the advantage of a facing crossover. As the train drew up well back from the inner home signal, the fireman of the pilot engine would be on the bottom cab step and, as it stopped, he'd be down between the two engines, slacking the coupling, splitting the vacuum bags and pushing them back on to their plugs. The signalman would have the road set, and as the pilot engine shot over the crossover, the driver of the train engine would have made his brake and be moving gently towards the signal. By the time the pilot engine was in the stub of the Ballyclare branch the road was pulled, and the train was away within two minutes of stopping. Such was the speed of operation the NCC inculcated into its men.

Moving on down the line, the other major gradient was the sharp and curving climb from the platform end at Ballymena up to Galgorm crossing. As related earlier in the book, this was one case where the train engine took the tablet, while the banker collected the banking staff which locked the instruments at Ballymena and Cullybackey.

In the up direction there was that steep climb at 1:75 out of Portrush, particularly difficult with a heavy train on a wet day. On the other hand, Portrush station was surrounded by sandhills, and while the sand which blew about the track was trying for signalmen who had to keep point blades clear, it certainly helped enginemen with adhesion! Since there was no bank staff arrangement, banking out of Portrush was (in theory anyway) prohibited, but at busy times in the days of the 4-4-0s it was not unknown for the second engine in a group of specials to push the first one out to Dhu Varren, then the third to assist the engine of the second, and so on.

One Saturday, GNR Q class No 134 arrived with a Sunday School special off the Great Northern, and was later booked to head the homeward procession of specials. An interested, and rather ribald, audience of NCC men – most of them secretly hoping to see the Great Northern engine come to grief – were reduced to silence as she stormed out of Portrush without any assistance at all. What they didn't know, of course, was that the same engine spent most of her life plodding up and down Carrickmore bank on the difficult Great Northern 'Derry Road'.

Billy McAfee, later to become CME at York Road, was a junior official in the traffic department in 1939 when he was sent to Portrush to supervise a busy night of Fireworks Specials. Having seen them all away, he travelled with the driver of the final train. He willingly accepted the driver's offer of his seat – no one turns down a chance to drive – but only when he got the guard's 'right away' did it dawn on him that, being the last special, they had no engine to assist them. McAfee was in a quandary: if he changed his mind, the story that "McAfee was afeard to take her out of Portrush" would be in Belfast ahead of the train. But if he stalled the train, so would the word that "McAfee cannae drive"! So with the darkness of the night covering his worried expression, he took a deep breath, wound the engine well out on the lever and put her on to the main valve. They stormed up past Glenmanus Siding with more pyrotechnics coming out of the engine than the firework display behind them at Ramore Head, and a well pleased (and relieved) McAfee got down at Coleraine to continue his journey home 'on the cushions'. But, as he remarked to me thirty years later, you just can't win. Next day, the breaking news at York Road was: "Thon McAfee can drive … but, oh, he's desperate hard on an engine"!

A later story, when McAfee had attained much more senior management status, centres on Coleraine. He was sent down to investigate some irregularity, and for good measure, as he stepped off the 8.35am at Coleraine he could see that mogul No 95 had got her tender off the road at the turntable. Slipping incognito (or so he thought) over to the shed, he heard the phone ringing in 'The Tarry'. Stepping into the empty room he lifted the receiver to hear an agitated voice from the barrier saying; "Mind yer eye, thon big b***d's on his way over". Judge the caller's consternation when he heard an all too familiar voice at the other end saying: "No, he's arrived"! In fact, McAfee was widely respected as an excellent railwayman, who unfortunately came to power at York Road just as the diesel era was beginning, and thus had no chance to make his mark as a steam engineer. He did much to help the fledgling RPSI through its early years, and – in a manner now almost impossible with modern railway management – he had the authority and experience to make instant decisions.

Longer and no less taxing was the southbound climb out of Ballymoney, a curving seven miles past Ballyboyland and up to the 'Twin Arches' at milepost 48½, with a short length of 1:87 for good measure. The story is told of a BNCR driver in bygone days, sitting impassively by his fireside while his wife was in labour upstairs. To the midwife's plea for his help, as her patient was having a hard time, he shouted up the stairs: "She doesn't know what a hard time is – she never had to take 48 wagons up Ballyboyland".

The third major climb facing a train from Portrush to Belfast was the long pull from Antrim up to Kingsbog Junction. Once over the top, that steep descent through Mossley had to be carefully watched. On one occasion, the driver of mogul No 101 brought a heavy goods over Kingsbog and, somewhere around Mossley, he stooped down to the firebox to get a light for his pipe. In the few seconds that his attention was distracted, the train had taken control of the engine, and with showers of sparks coming off all the brakeblocks, they came over the viaduct nearly as fast as an up express. It was just as well they had a clear road into the goods yard at Belfast, for they'd never have stopped and, when the shaken crew brought 101 off the train, the engine brakeblocks were so badly worn that they needed to use the tender handbrake to stop her at the shed!

This story reminded me of Frank Dunlop assessing a young fireman many years later. With something like the above experience in mind, the Inspector wisely remarked: "We've taught him how to run, so now we'll have to teach him how to stop!" Coincidentally, the same young man had fired in the early 1960s to Ned Nelson, who in his own young day was the fireman on 2-4-0 No 56 on a Cookstown train in the 1930s with no less that OS Nock on the footplate.

On the minor lines, the Ballyclare branch had a sharp climb up to Kingsbog Junction. There was a famous occasion during World War Two when a train of munitions and bren-gun carriers stuck near the Junction on a night of thunder, lightning and torrential rain. It took the combined efforts of Scotch engines Nos 70 and 71, as well as the engine off the night goods from Coleraine and another mogul, to get the train on the move.

The Dungiven line, with a 5½ mile climb varying from 1:90 to 1:70 near Ardmore, also had a reputation for difficulty, but with light loads and an A class 4-4-0 blowing off at 200psi there were few problems. One morning during the war, the crew of the morning goods left their engine in charge of the signalman whilst they went off up the town in search of food to supplement their ration book entitlement. In return for his unofficial assistance, they agreed to the signalman's request for "a barrow of coal". When they got back, the fireman climbed into the tender, immediately shouting down to the driver: "It's all gone!" Fortunately this was an exaggeration, but since the signalman's numerous friends had included themselves in the crew's invitation, the remaining coal had to be eked out with broken sleepers and scrap timber to get the train back to Derry!

The Narrow Gauge

Whereas the broad gauge gradients were not generally excessive, those on the narrow gauge were quite extreme. Leaving Ballycastle, for example, the line climbed for three miles to Capecastle at 1:60 to 1:80, and in the other direction there was a pull of 1:80 out of Ballymoney. That descent to sea level at Ballycastle was the scene of a spectacular runaway in 1945 when 2-4-2T No 41 got away on the Ballylig curves at milepost 13, just after Capecastle. The engine thundered over the Tow Viaduct at perhaps 40mph, smashed down two fences at the station, and crossed the road into a stream, where she lay for the following week. Driver McKissock stayed with the engine, but fireman Heffron jumped off as they ran up the platform. Fortunately, neither was seriously injured. A second runaway took place on a frosty January night in 1949, when No 44 slid through Ballycastle station and once more ended up on the other side of the road, fortunately without injury to the crew of McDuff and McKinley.

If the Ballycastle banks were hard, those on the Ballymena lines were fearsome. The Parkmore line climbed at 1:37, 1:39 and 1:60 to Essathohan siding before descending at 1:40 and 1:37 to Retreat. It was as well the line finished here, as the projected extension to Red Bay on the coast would have entailed a bank of 1:21, or 1:31 with South African style reverses.

On the Larne line, there was a climb of 1:36 out of the town, and the Inver bank was a worry in both directions. Drivers of trains going into Larne were cautioned to have full command of their brakes, and to stop if necessary to pin down brakes. At the other end of the line, the final eight miles to Ballynashee were at 1:87 and 1:75. Averaging 25mph may not seem very fast, but for the Boat Express, worked by a 2-4-2 tank, to keep the schedule over these gradients required some of the fastest work of any Irish narrow gauge line.

Few records exist of locomotive work on the Ballymena line, though Billy Hanley told me that the compounds could run easily into the thirties, and climbed well with modest loads. Both Billy and Davie McDonald confirmed that the small grates of these engines required careful work on the fireman's part. 'The Bruiser', that giant 2-6-0 tank already described, was well liked by drivers as its small coupled wheels made it sure-footed on the steep gradients. By contrast, the two Ballycastle 4-4-2 tanks which were used on the Ballymena & Larne for a while were disliked for their propensity to slip.

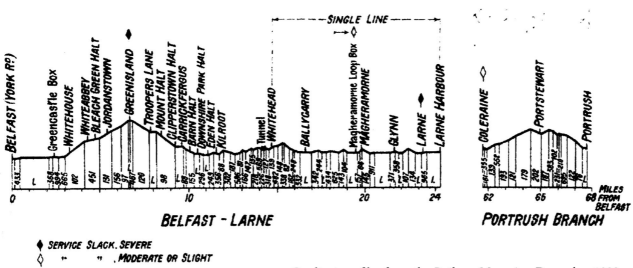

Gradient profiles from the *Railway Magazine*, December 1930

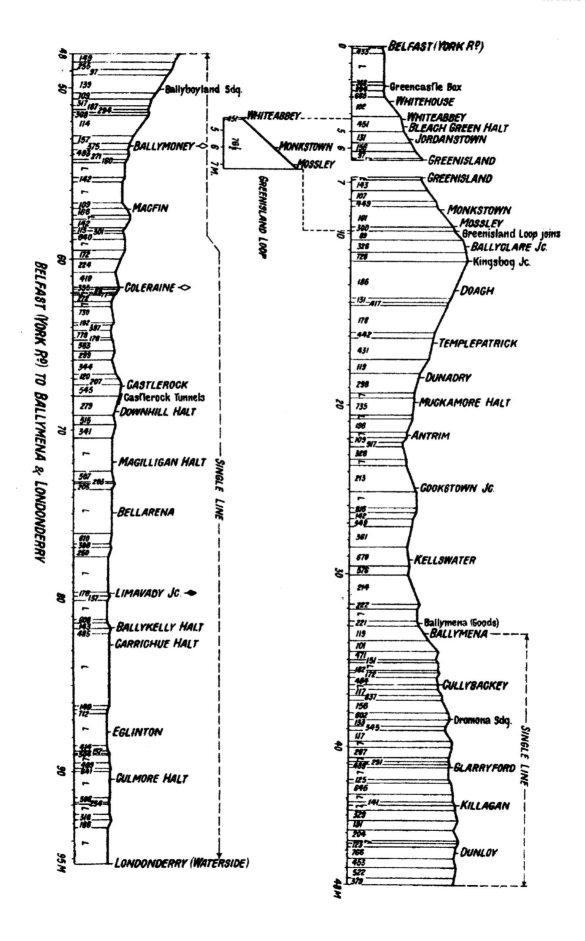

Appendix 4
Locomotive Weight Distribution Diagrams

Important note to modellers

These NCC weight-distribution diagrams were for use by the Civil Engineering Department and have been photocoped at various times from originals or A3 copies of the originals. The originals were to a large scale and were reduced to A4 to facilitate storage and preservation. Because this was done by various people, and at different times, they will not be to a constant scale. In any case it would be impossible to reproduce them at a convenient 4mm to the foot scale on this page size without placing them at right angles to the page.

Being a modeller myself, I am acutely aware of the inconvenience this may create but, since the principal dimensions are printed on the drawings, it should be possible to determine the scale without a degree in Mathematics! Most NCC passenger engines conveniently have a driving wheel diameter of six feet, which is a good starting point.

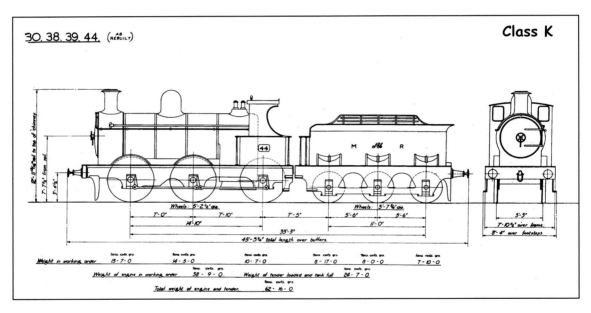

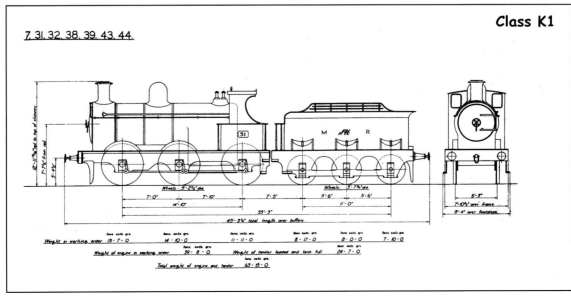

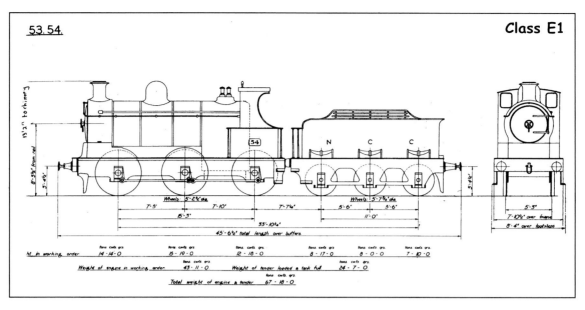

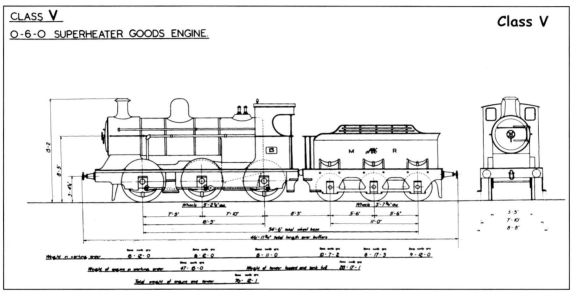

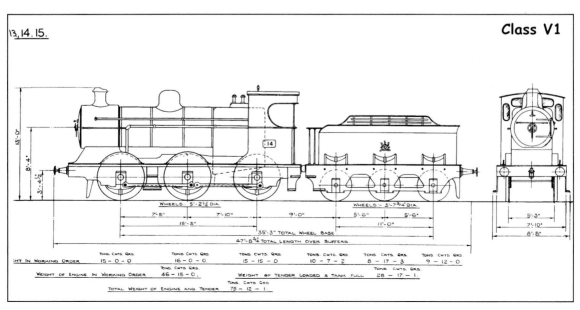

Class J

25. 47. 48. 49.

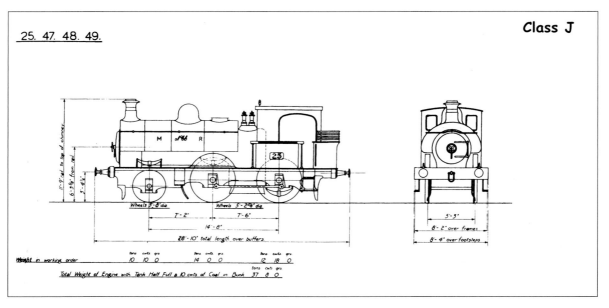

Class C

21. 33. 50. 51. 52. 56. 57

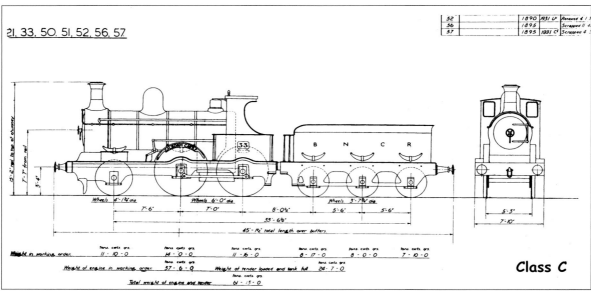

Class B

24. 59. 60. 61. 62.

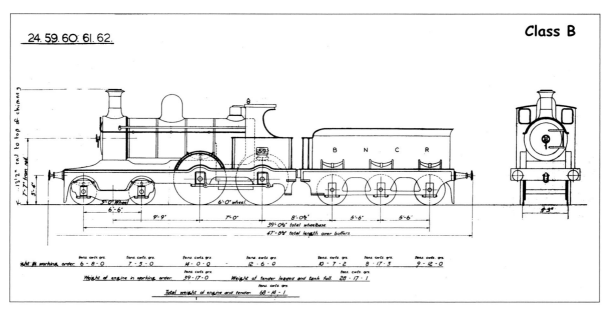

4, 5, 9, 17, 20, 34, 63, 64, 65, 66, 67, 68.

20		1905	1929	U²	Renewed 64 12.2
34	Queen Alexandra	1901	1929	A¹	Scrapped 10.4
63		1905	1936	U²	Renewed 67 5.34
64		1905	1929	A¹	Scrapped 8.54
65		1905	1929	A¹	Scrapped 10.54
66		1905	1930	A¹	Scrapped 8.52
67		1908	1934	U²	Renewed 65 5.34
68		1908	1927	A¹	Scrapped 11.47

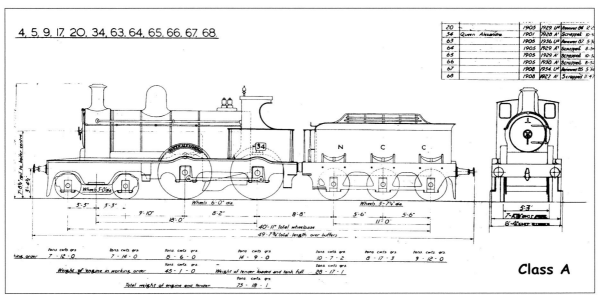

Class A

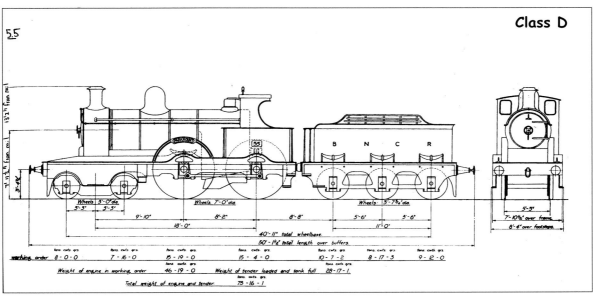

Class D

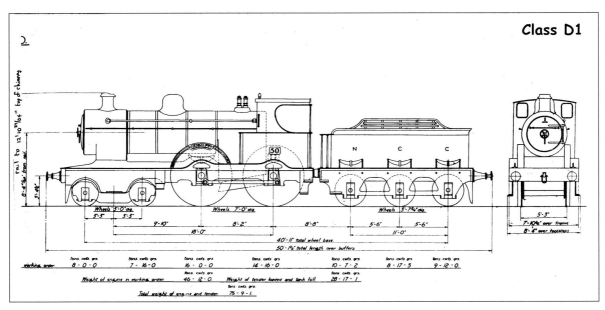

Class D1

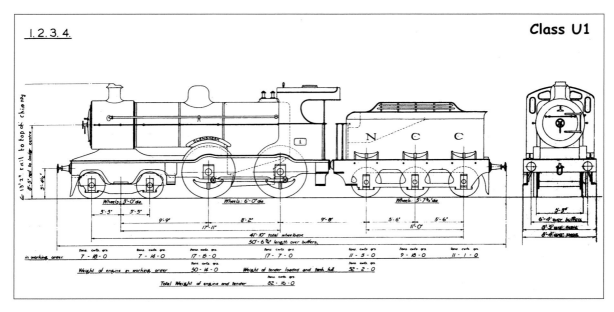

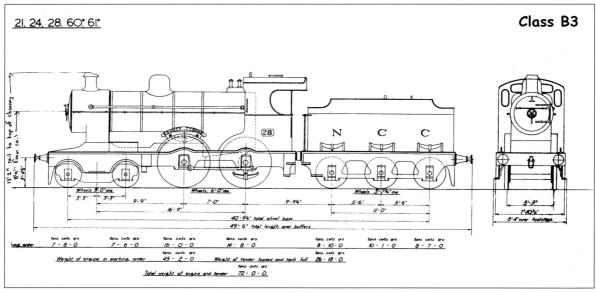

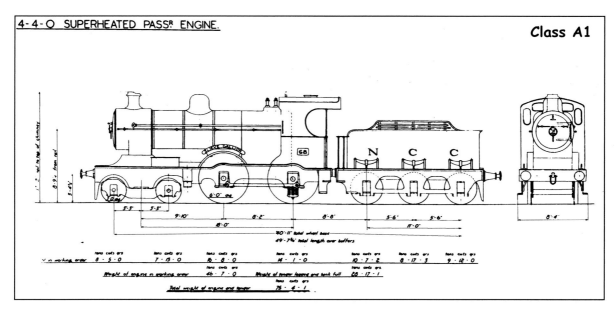

4-4-0 SUPERHEATED PASSR ENGINE.

Class U2

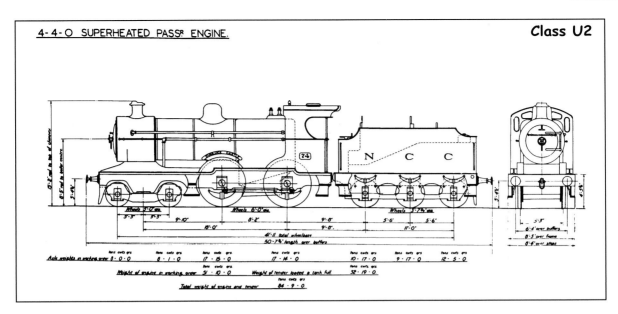

70, 71, 72, 73, 84, 85, 86, 87.

Class U2

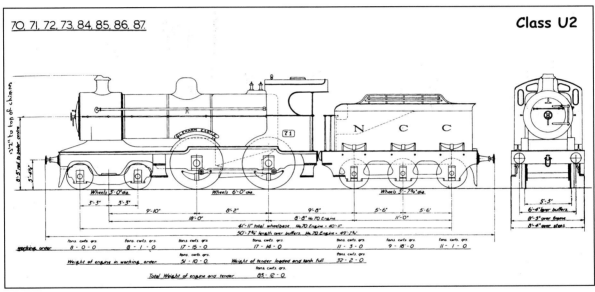

CLASS N.

O-4-O SADDLE TANK SHUNTING ENGINE.

Class N

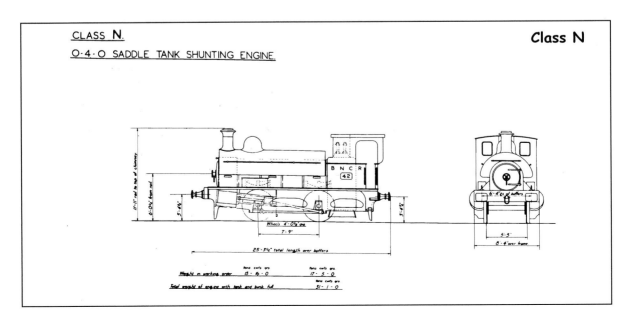

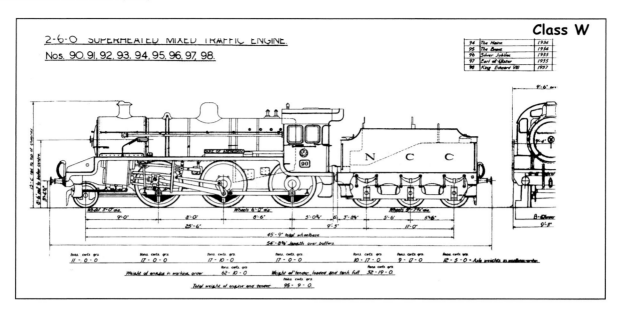

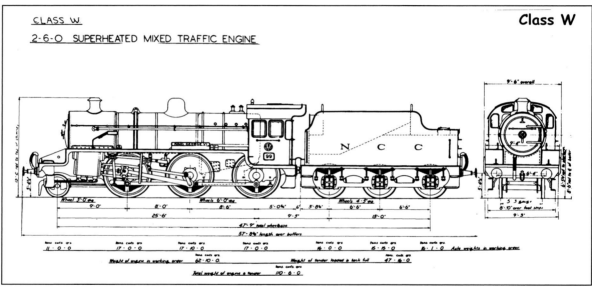

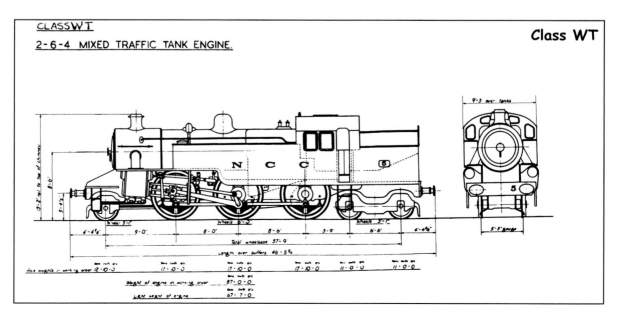

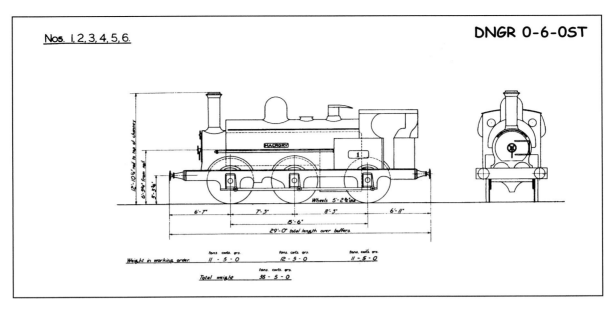

Nos. 1, 2, 3, 4, 5, 6.

DNGR 0-6-0ST

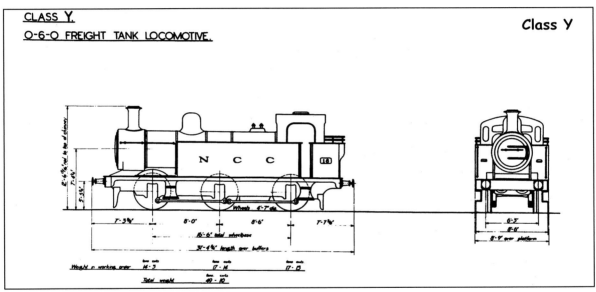

CLASS Y.

0-6-0 FREIGHT TANK LOCOMOTIVE.

Class Y

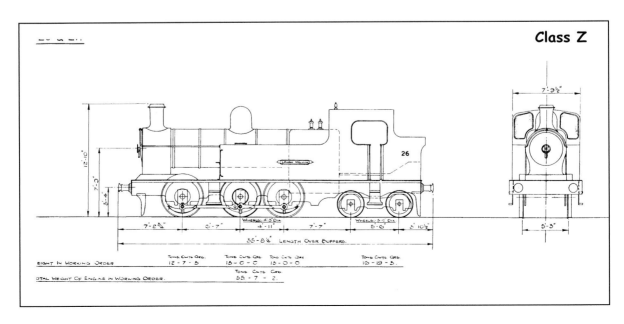

Class Z

Class S

43, 44.

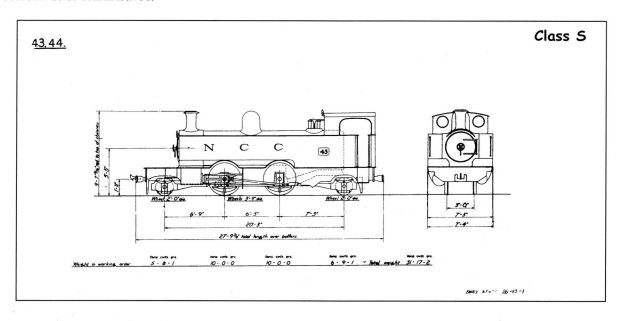

Class S1

CLASS S¹

2-4-2 COMPOUND SIDE TANK ENGINE.

No's. 41, 42.

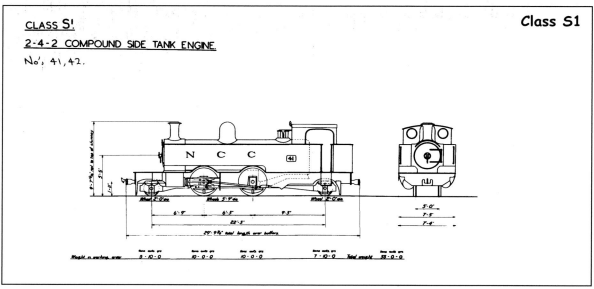

Class S2

No. 110.

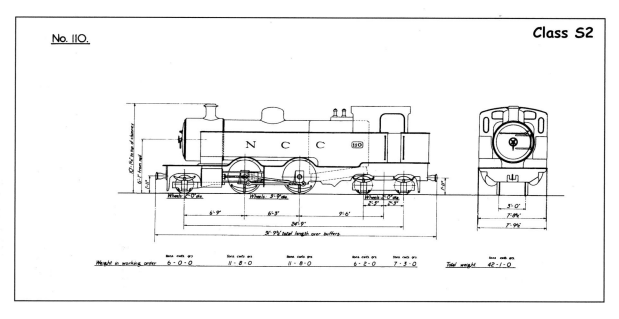

Class T

113. 114.

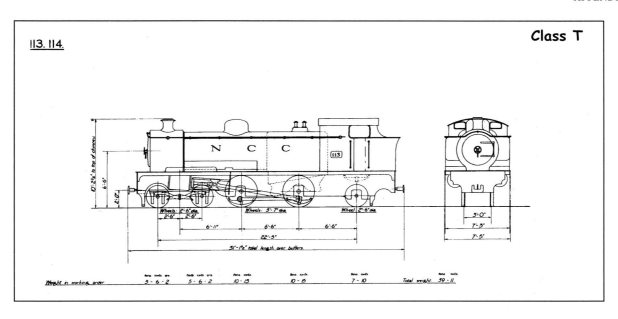

Class X, No 17

CLASS X

0-6-0 DIESEL MECHANICAL SHUNTING ENGINE.

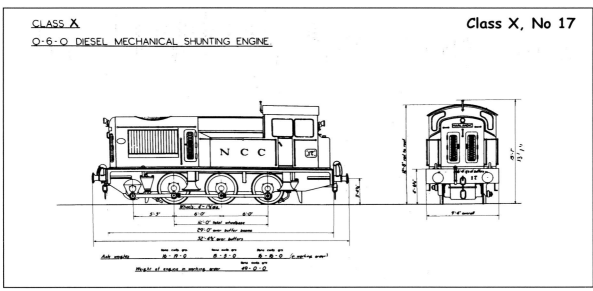

No 22

0-6-0 DIESEL MECHANICAL SHUNTING ENGINE.

No. 22.

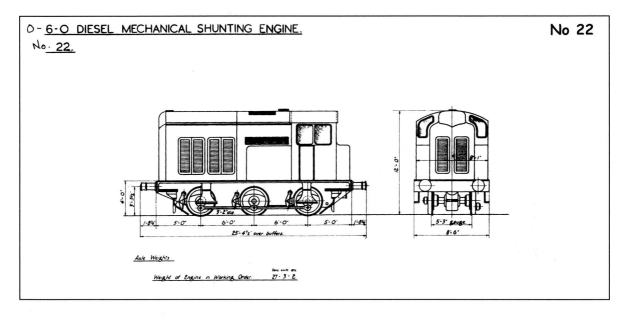

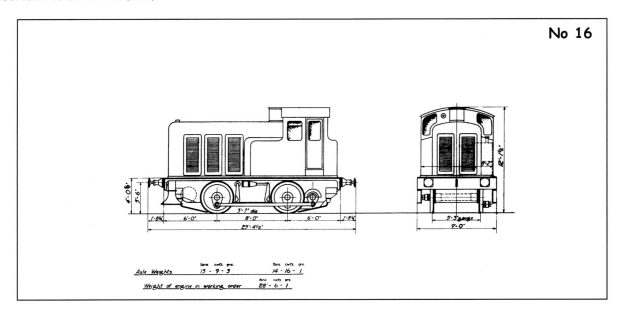

No 16

Axle Weights — 13 - 9 - 3 | 14 - 16 - 1

Weight of engine in working order — 28 - 6 - 1

2·2·0 + 0·2·2 DIESEL ELECTRIC LOCOMOTIVE (PASS^R)

No 28

No. 28.

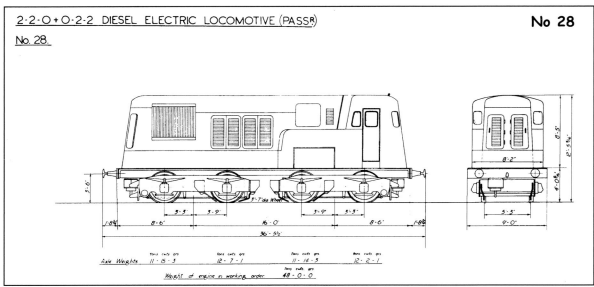

Axle Weights — 11 - 15 - 3 | 12 - 7 - 1 | 11 - 14 - 3 | 12 - 2 - 1

Weight of engine in working order — 48 - 0 - 0

No 1

54 FOOT **DIESEL** RAILCAR.

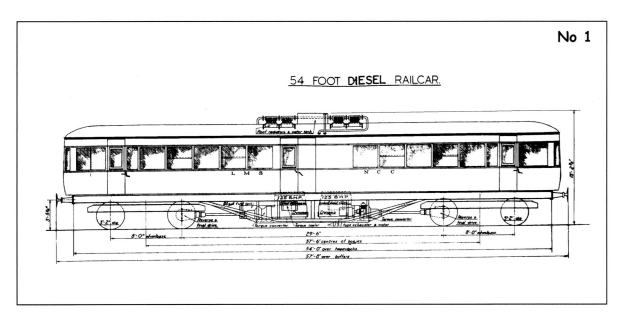

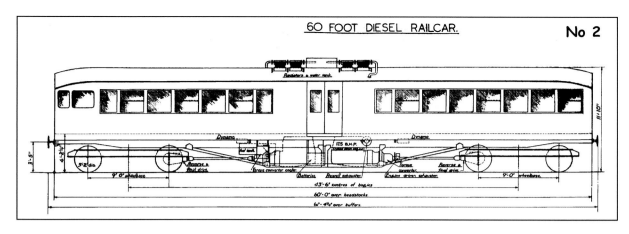

60 FOOT DIESEL RAILCAR.   No 2

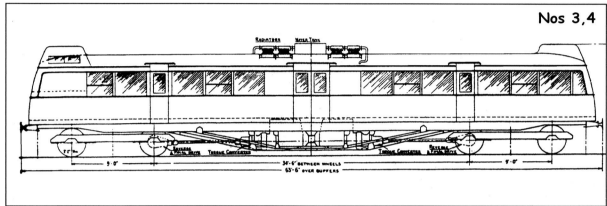

Nos 3,4

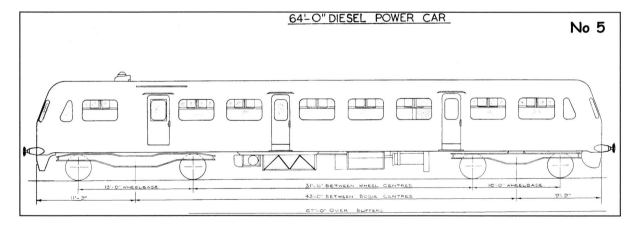

64'-0" DIESEL POWER CAR   No 5

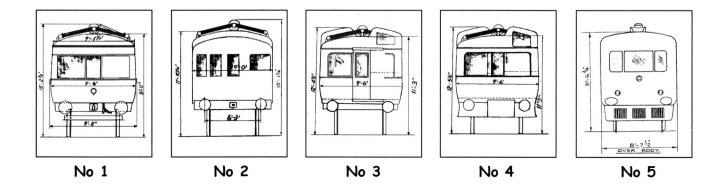

No 1   No 2   No 3   No 4   No 5

Appendix 5
NCC Maximum Load Tables 1896 and 1938

1896

1896

BROAD GAUGE.											
INCLINES.	Class A. 1, 2, 3, 4, 34.		Class B. 5, 12, 13, 14, 15, 16, 17, 20, 25, 47, 48, 49.		Class C. 6,8,9,10,11,21,22, 23, 27, 29, 33, 40, 41, 45, 46, 50, 51, 52, 55, 56, 57, 58.		Class D. 7, 18, 19, 28, 30, 31, 32, 35, 36, 37, 38,39,43,44,53,54		Class E. 26.			
	No. of Vehicles.		No. of Vehicles.		No. of Vehicles.		No. of Vehicles.		No. of Vehicles.			
	Passgrs.	Goods.	Passgrs.	Goods.	Passgrs.	Goods.	Passgrs.	Goods.	Passgrs.	Goods.		
DOWN.												
Belfast to Greenisland	12	18	13	20	14	22	18	25	17	23	...	...
Carrick „ Larne	12	18	13	20	14	22	18	25	17	23	...	...
Greenisland „ Ballyclare Jun.	12	15	13	17	14	19	18	25	17	23	...	...
Ballymena „ Cullybackey	12	17	13	19	14	21	18	28	17	26	...	...
C.town Jun. „ Toome	14	20	15	22	16	24	18	35	17	33	...	...
Toome „ Cookstown	13	18	14	20	15	22	18	30	17	28	...	...
Limavady „ Dungiven	9	9	10	11	11	13	15	18	15	17	...	...
Magherafelt „ Draperstown	9	9	10	11	11	13	15	18	15	17	...	...
Magherafelt „ Coleraine	10	14	11	15	12	18	15	20	15	19	...	...
Coleraine „ Portrush	14	15	15	17	16	19	18	25	17	23	...	...
UP.												
Larne to Carrick	12	18	13	20	14	22	18	28	17	26	...	...
Carrick „ Greenisland	12	15	13	17	14	19	18	25	17	23	...	...
Ballyclare „ Ballyclare Jun.	10	14	11	16	12	18	15	18	15	17	...	...
Dunadry „ Ballyclare Jun.	13	17	14	19	15	21	18	30	17	28	...	...
Ballymoney „ Killagan	13	17	14	19	15	21	18	30	17	28	...	...
Castlerock „ Coleraine	14	20	15	22	16	24	18	30	17	28	...	...
Cookstown „ Cookstown Jun.	15	22	16	24	17	26	18	35	17	33	...	...
Dungiven „ Limavady	9	9	10	11	11	13	15	18	15	17	...	...
Coleraine „ Magherafelt	10	14	11	16	12	18	15	20	15	18	...	...
Portrush „ Coleraine	14	15	15	17	16	19	18	25	17	23	...	...

NARROW GAUGE.	Class A. 60, 61, 62, 63, 64.	Class B. 65, 66, 67.	Class C. 68, 69, 70.
Larne to Kilwaughter	6 Wagons and Van.	10 Wagons and Van.	12 Wagons and Vans.
Kilwaughter „ Ballyboley	10 „ „ „	15 „ „ „	17 „ „ „
Ballyboley „ Ballynashee	11 „ „ „	16 „ „ „	18 „ „ „
Ballynashee „ Ballymena	20 „ „ „	20 „ „ „	20 „ „ „
Ballymena „ Larne	9 „ „ „	14 „ „ „	16 „ „ „
Parkmore „ Knockanally	16 „ „ „	16 „ „ „	16 „ „ „
Knockanally „ Ballymena	24 „ „ „	24 „ „ „	24 „ „ „
Iron Ore Sidings from Mines	16 „ „ „	16 „ „ „	16 „ „ „
Ballymena to Knockanally	24 „ „ „ } Empty	24 „ „ „ } Empty	24 „ „ „ } Empt
Knockanally „ Parkmore	16 „ „ „	16 „ „ „	16 „ „ „
Iron Ore Sidings to Mines	16 „ „ „	16 „ „ „	16 „ „ „

The loads given above are for the guidance of Enginemen and Traffic Officials, and refer in all cases—except where otherwise mentioned—to fully loaded standard vehicles, and they are such as will permit the Engine to stop and start from any Station or Siding without danger of stalling under ordinary circumstances. Under exceptional circumstances, or when an Engine from any reason is unable to work its proper load, the Driver must decide how many vehicles it can work, and his decision is to be final.

When the vehicles forming a train are not fully loaded, or when a train is timed not to stop at Stations situated at or near inclines, the driver must use his discretion as to what additional vehicles he may take, care being taken not to overload the Engine as to run the danger of stalling.

In calculating the number of vehicles in a train, the following rules must be observed:—

Each Long Broad Gauge Passenger Vehicle will count as 2 Loaded Wagons in the case of mixed trains.
„ Short Broad Gauge Passenger „ „ „ „ 1 Loaded Wagon „ „ „ „
„ Long Narrow Gauge Carriage „ „ „ 2 „ Wagons.
„ Short „ „ „ „ „ „ 1 „ Wagon.
„ Loaded 16 ton Ore Wagon „ „ „ 2 „ Wagons
Two „ 8 ton N.G. Hopper Wagons „ „ „ 3 „ „
Three Broad Gauge Empty Wagons „ „ „ 2 „ „
Two Narrow Gauge „ „ „ „ „ 1 „ Wagon.

1938

Classification of Engines according to their power and shewing the load (Brake Vans included) which can be worked by each Engine up the inclines on the L. M. & S. R. (N.C.C.) System.

BROAD GAUGE.

INCLINES		CLASS A. 23, 45, 46, 51, 55, 56, 57.		CLASS B 21, 24, 28, 30, 31, 33, 34, 43, 44, 50, 54, 58, 60, 61, 62, 64, 65, 66, 68, 69.		CLASS C. 1, 2, 3, 4, 70, 71, 72, 73, 74, 75, 76, 77, 78, 79, 80, 81, 82, 83, 84, 85, 86, 87.		CLASS D. 13, 14, 15.		CLASS E. 90, 91, 92, 93, 94, 95, 96, 97, 98, 99.	
		Gross Load in Tons.		Gross Load in Tons.		Gross Load in Tons.		Gross Load in Tons.		Gross Load in Tons.	
		Passr.	Goods.	Passr.	Goods.	Passr.	Goods.	Passr.	Goods.	Passr.	Goods.
DOWN.											
Belfast	to Greenisland	205	260	290	350	325	400	340	430	420	530
Carrickfergus	„ Larne	205	260	290	350	325	400	340	430	420	530
Greenisland	„ Ballyclare Jct.	135	220	235	280	255	320	300	370	330	450
Belfast	„ Ballyclare Jct.	135	220	235	280	255	320	300	370	330	450
Ballymena	„ Cullybackey	205	260	290	350	325	400	340	430	420	530
Cookstown Jct.	„ Toome	205	260	290	350	325	400	340	430	420	530
Toome	„ Cookstown	155	195	240	300	270	340	290	350	360	470
Limavady	„ Dungiven	155	195	240	300	270	340	290	350	360	470
Magherafelt	„ Draperstown	155	195	240	300	270	340	290	350	360	470
„	„ Coleraine	155	195	240	300	270	340	290	350	360	470
Coleraine	„ Portrush	205	260	290	350	325	400	340	430	420	530
UP.											
Larne	to Carrickfergus	185	235	270	340	305	375	325	405	400	500
Carrickfergus	„ Greenisland	185	235	270	340	305	375	325	405	400	500
Ballyclare	„ Ballyclare Jct.	155	195	240	300	270	340	290	350	360	470
Dunadry	„ „	205	260	290	350	325	400	340	430	420	530
Ballymoney	„ Killagan	185	235	270	340	305	375	325	405	400	500
Castlerock	„ Coleraine	205	260	290	350	325	400	340	430	420	530
Cookstown	„ Cookstown Jct.	205	260	290	350	325	400	340	430	420	530
Dungiven	„ Limavady	155	195	240	300	270	340	290	350	360	470
Coleraine	„ Magherafelt	185	235	270	340	305	375	325	405	400	500
Portrush	„ Coleraine	185	235	270	340	305	375	325	405	400	500

The loads given above are for the guidance of Enginemen and Traffic Officials, and they are such as will permit the Engine to stop at and start from any Station or Siding without danger of stalling under ordinary circumstances. Under exceptional circumstances or when an Engine from any reason is unable to work its proper load, the Driver must decide what tonnage it can work, and his decision is to be final.

When a train is timed not to stop at Stations situated at or near inclines, the Driver must use his discretion as to what additional tonnage he may take, care being taken not to so overload the Engine as to run the danger of stalling.

	Loaded with Grain or other Bag Stuffs.	Bricks, Slates, Coal or other Minerals.	Manure in Bulk.	Timber.	Sundries.	Cattle.	Empty.
Gross Loads of Goods Trains to be calculated thus :—	Tons.	Tons.	Tons.	Tons.	Tons.	Tons.	Tons.
6-Wheel Wagons	22	20	—	20	15	—	7
4-Wheel Ordinary Open Wagon	12½	13	15	12	8	—	5
15 Ton Wagon	23	22	—	—	18	—	8
20 Ton Wagon	29	28	—	—	19	—	9
Bogie Wagon	42	—	42	—	27	—	13
Covered Wagon	13	—	—	—	9	—	6
Cattle Wagon	—	—	—	—	—	9	5
Timber Truck	10	—	—	10	—	—	10
Brake Van	—	—	—	—	12	—	10
„ „	—	—	—	—	23	—	20

NARROW GAUGE.

	Class A. 101, 102, 103, 104, 110, 111, 113, 114.
Larne to Kilwaughter	11 Wagons and Van.
Kilwaughter „ Ballyboley	17 „ „ „
Ballyboley „ Ballynashee	18 „ „ „
Ballynashee „ Ballymena	20 „ „ „
Ballymena „ Larne	16 „ „ „
Rathkenny „ Ballymena	24 „ „ „
Ballymena „ Rathkenny	24 „ „ „
Ballymoney to Ballycastle	15 „ „ „
Ballycastle to Capecastle	14 „ „ „
Capecastle to Ballymoney	15 „ „ „

Each Long Narrow Gauge Carriage will count as 2 Loaded Wagons.

Two Loaded 8 ton N.G. Hopper Wagons will count as 3 Loaded Wagons.

Two Narrow Gauge Empty Wagons will count as 1 Loaded Wagon.

7 ton end-tip wagons. Between any two points loaded 7 ton end-tip wagons to be counted as follows: the minimum number to be taken into consideration being 3.

3 equal 4
4 equal 5
5 equal 6
6 equal 8
and so on.

Bibliography

Liddle, LH, *Steam Finale* (IRRS London Area 1964)

Hills, RL and Patrick, D, *Beyer Peacock – Locomotive Builders to the World* (TPC 1982)

Lloyd, Joseph, *Beyer Peacock drawings*

Nock, OS, *Irish Steam* (David and Charles 1982)

Johnston, Norman, 'The LMS Ancestry of No 4' (In *Five Foot Three* No xx 1976)

Arnold, RM, *NCC Saga* (Blackstaff Press 1973)

Arnold, RM, *Steam Over Belfast Lough* (RPSI 1969)

Ahrons, RM, *Locomotive and Train Working in the Latter Part of the Nineteenth Century*, Vol 6 (Heffer 1954)

Journal of the Irish Railway Record Society (various issues)

Five Foot Three, the journal of the Railway Preservation Society of Ireland (various issues)

Houston, JH, Notes on NCC locomotives

Clements, RM, Notes on NCC locomotives (Irish Railway Record Society archives)

BELFAST & BALLYMENA RAILWAY.

HOURS OF DEPARTURE ON AND AFTER 1st JULY, 1848.

A timetable poster showing DOWN TRAINS FROM BELFAST, CARRICKFERGUS, AND RANDALSTOWN and UP TRAINS FROM BALLYMENA, RANDALSTOWN, AND CARRICKFERGUS, with Week Days, Sundays, and Fares columns, followed by General Notices. Signed THOMAS H. HIGGIN, General Manager.

Index

Other Steam Locomotives

Diesel Locomotives

Railcars